Kewda

About the Authors

Dr. P. N. Jagadev, presently working as Professor, Department of Plant Breeding and Genetics, Odisha University of Agriculture and Technology, Bhubaneswar has obtained his B. Sc.(Ag.) and M. Sc. (Ag.) degrees from OUAT, Bhubaneswar, Ph. D. degree from GBPUAT, Pantnagar and Post-Doc. Training from DDPSC, Washington University, St. Louis, USA. He also served as the Dean of Research, Dean of Extension Education, Professor and Head, Department of Bioinformatics of OUAT, Bhubaneswar, Odisha,. He has research areas of specialization in plant stress breeding, biotechnology and bioinformatics with more than 33 years of distinguished career on teaching, research and extension. He travelled many countries like USA, Australia, South Korea, Hong Kong, Thailand and Kenya on different R&D assignments. He has guided 55 students of P.G., Ph.D. and PDF levels. To his credit, he has 258 publications (research papers-86, conference/ symposium papers-40, books/manuals-33, book chapters-9, technical booklets-25 and popular scientific articles-65). He is a Fellow of Indian Society of Genetics and Plant Breeding. He has also developed seven crop varieties on rice, sugarcane, maize, cotton, jute and mesta for the farmers of Odisha and other states. He has the distinction of receiving several awards like Krutti Krushi Vaigyanik Samman and Best Agricultural Scientist Award from the Government of Odisha, Krushi Ratna Award from Krushak Sambad and Best Director Award from Education Expo, TV, Mumbai

Dr. Sashikala Beura, presently working as the Professor and Head, Department of Floriculture and Landscaping and as Director, Biotechnology-cum-Tissue Culture Centre, Odisha University of Agriculture and Technology, Bhubaneswar, has obtained her M.Sc. (Ag.) degree in Horticulture from OUAT, Bhubaneswar with University Gold Medal and Ph.D. degree from GBPUAT, Pantnagar with CSIR Senior Research Fellowship. She has research areas of specialization in Floriculture, Landscaping, Biotechnology and Bioinformatics with more than 28 years of distinguished career on teaching, research and extension. She travelled many countries like Australia, Thailand and Iran on different R&D assignments. She has guided 33 PG and 3 Ph.D. students. To her credit, she has 255 publications (research papers-79, conference/symposium papers-78, books/manuals-8, book chapters-9, technical booklets-15 and popular scientific articles-66). She has the distinction of receiving several awards like Alaka Manjari Rath Best Woman Scientist Award from the Governor of Odisha, Best Faculty Award from Education Expo TV, Kolkata and two best paper awards. She has also developed a cashewnut hybrid 'Jagannath' and introduced the protected cultivation of gerbera, lilium and anthurium for the farmers of Odisha.

Dr. Trinath Maharana, former Professor and Head, Department of Horticulture, Odisha University of Agriculture and Technology, Bhubaneswar and Emeritus Professor, ICAR has obtained his M. Sc. (Ag.) and Ph.D. degrees from OUAT, Bhubaneswar. He also served as the Associate Director of Research (Seeds), Directorate of Research, OUAT, Bhubaneswar. He is an expert of tribal horticulture and tuber crops and well known for his excellence in academics and research ideas with more than 35 years of distinguished career on teaching, research and extension. He was also the pioneer to start research on floriculture and tuber crops in OUAT about five decades back. He worked as the Principal Investigator of several adhoc projects on kewda, minor cucurbits, medicinal and aromatic plants etc. He also guided 50 students of M. Sc. (Ag.) and Ph. D. levels and published more than 75 research papers in journals of national/ international repute. Besides, he has authored a number of text books in Odia and served as QRT member (ICAR) of palms, cashewnut and betelvine. He was also the Secretary of the Odisha Horticultural Society and Association of Agricultural Scientist, Bhubaneswar and presently, he is the President of Odisha Horticultural Society.

Kewda
Cultivation and Perfume Production

P.N. Jagadev
Professor (Plant Breeding & Genetics) and
Former Dean of Research and Extension Education
Odisha University of Agriculture and Technology
Bhubaneswar-751003, Odisha

Sashikala Beura
Professor & Head (Floriculture & Landscaping) and
Director, Biotechnology-cum-Tissue Culture Centre
Odisha University of Agriculture and Technology
Bhubaneswar-751003, Odisha

Trinath Maharana
Former Professor and Head (Horticulture) and
Emeritus Professor (ICAR)
Odisha University of Agriculture and Technology
Bhubaneswar-751003, Odisha

nipa®
NEW INDIA PUBLISHING AGENCY
New Delhi – 110 034

NEW INDIA PUBLISHING AGENCY
101, Vikas Surya Plaza, CU Block, LSC Market
Pitam Pura, New Delhi 110 034, India
Phone: + 91 (11) 27 34 17 17 Fax: + 91 (11) 27 34 16 16
Email: info@nipabooks.com
Web: www.nipabooks.com

Feedback at feedbacks@nipabooks.com

ISBN: 978-93-89571-73-8

Composed and Designed by NIPA

Dedicated to

The Kewda Growers of Ganjam District, Odisha

सत्यमेव जयते

त्रिलोचन महापात्र, पी.एच.डी.
एफ.एन.ए., एफ.एन.ए.एस.सी., एफ.एन.ए.ए.एस
सचिव एवं महानिदेशक

TRILOCHAN MOHAPATRA, Ph.D.
FNA, FNASc, FNAAS
SECRETARY & DIRECTOR GENERAL

भारत सरकार
कृषि अनुसंधान और शिक्षा विभाग एवं
भारतीय कृषि अनुसंधान परिषद
कृषि एवं किसान कल्याण मंत्रालय, कृषि भवन, नई दिल्ली 110 001

GOVERNMENT OF INDIA
DEPARTMENT OF AGRICULTURAL RESEARCH & EDUCATION
AND
INDIAN COUNCIL OF AGRICULTURAL RESEARCH
MINISTRY OF AGRICULTURE AND FARMERS WELFARE
KRISHI BHAVAN, NEW DELHI 110 001
Tel. : 23382629; 23386711 Fax : 91-11-23384773
E-mail : dg.icar@nic.in

Foreword

Kewda (*Pandanus fascicularis* Lam.), locally known as 'Kia' occurs widely in the coastal regions of India and distributed upto Central Malaysia. It is commercially important due to the utilization of its aromatic male spadices in the production of highly prized kewda perfume. The perfume has great demand in India and abroad, especially from Arab countries, earning more than 50 crores of rupees annually from export. It is also important in generating economic man-days by utilizing the vast areas of coastal waste saline land. The kewda distillation industry has a great potential for expansion by exploiting the overloaded natural plantations along the Odisha coast. Hence, this crop has a tremendous scope in ensuring livelihood for rural people through industrial scale of production and to bring high profit for the perfumery industry.

I sincerely appreciate the efforts of Dr. P.N. Jagadev, Professor (Plant Breeding and Genetics) and Former Dean of Research and Extension Education and his team for bringing out a book on 'Kewda: Cultivation and Perfume Production'. This book incorporates various aspects related to Kewda such as origin and distribution, botany and nomenclature, species and varieties, breeding and crop improvement, soil and climate, propagation, cultural practices, harvesting and yield of flower, chemical composition, insect pests and diseases, production and marketing of perfumes, economics of cultivation and other uses of the crop. As brought to my knowledge the contents of this book are informative with a scope to take up further scientific research and develop entrepreneurship. This publication would be highly useful for the students, teachers and scientists of horticultural sciences especially of aromatic crops. Progressive farmers and traders concerned with cultivation and perfumery trade would also be equally benefitted from this book.

Dated the 26th November, 2019
New Delhi

Trilochan Mohapatra

Preface

The male spadices (grouped in flowers) of kewda (*Pandanus fascicularis* Lam.) are highly aromatic and collected for use in the preparation of kewda perfumes, which have a growing demand in domestic and international markets. Hence, the ultimate economic value of kewda plantations depends on the quality of flower production. The plant normally takes about five years after planting for flowering initiation. It is estimated that about 30 to 40 million flowers distilled annually in the Ganjam district of Odisha accounting for about 90 % of our country's total production of kewda perfumes mostly for kewda attar, water and oil. Besides kewda perfume industry, small cottage industries can also be set up to make various household usable items by utilizing different plant parts of kewda for the rural areas. The new regions of natural growing kewda plantations in good concentration should be identified for selection and propagation of superior strains to increase the flower productivity and a conscious cultivation of kewda deserves promotion by involving various agencies instead of leaving it to grow naturally. Therefore, the authors felt an imperative need to compile all the research findings and the collected information available through this book for development of appropriate cultural practices to boost high flower and perfume production in kewda.

The present book on 'Kewda' contains 15 chapters including some fundamental aspects and understandings of the crop on origin and distribution, botany, breeding and crop improvement, cultivation, insect pests and diseases, perfume production, marketing and economics of cultivation with various economic uses of the crop. The book is written in a simplified form to make it very useful for the students, farmers, traders and researchers engaged in medicinal and aromatic crops. The efforts have been made to present the book in a systematic and comprehensive manner with illustrative tables, plates and figures, wherever felt necessary.

We feel immense pleasure to express our heartfelt gratitude to the ICAR for sanctioning a project on *Pandanus* during the yester years, where the authors worked extensively and this book is an outcome of that.

We are highly grateful to Dr. T. Mohapatra, Secretary, DARE and DG, ICAR, Govt. of India for writing a foreword. We also like to acknowledge the support and encouragement received from Dr. R. K. Mohapatra, Former Chief Librarian, OUAT and appreciate the efforts of M/s New India Publishing Agency, New Delhi for publishing this book in a nice presentable form.

We wish to express our sincere gratitude to the kewda growers of Ganjam district, who contribute a lot of information through interactions of their secretive trade practices to make this publication, a success and as a respect, we dedicate this book to them. We further express our appreciation to Mr. A. Kar, P.A. to Vice-Chancellor, OUAT for typing the manuscript on the computer in a neatly manner within the shortest possible time.

Last but not the least, we extend profuse thanks to our family members for their constant encouragement and support all throughout in making our mission successful to bring this book in the present form.

Dated the 1st December, 2019

P. N. Jagadev
Sashikala Beura
T. Maharana

Contents

List of Figures

List of Plates

List of Tables

List of Colour Plates

1

Introduction

Among the vast resources of plants, aromatic and medicinal plants are highly beneficial to mankind for supplying medicines and essence. Olfactory gland bearing aromatic plants are few and many of them are located mainly in tropics and sub-tropics. Ecological conditions in the tropics are so conducive for such plants that out of 75,000 edible plants, majority are found in this region. (Chai *et al.*, 1979, Jong, 1979)

Among the aromatic plants, kewda (*Pandanus fascicularis* Lam.), a common species of screwpine, is commercially important for its highly aromatic male spadices used in the production of highly prized kewda perfumes. Kewda attar, water and oil (*rooh kewda*) are the important perfumery products. Kewda attar blends well with most of the fancy perfumes, in preparation of lotions, cosmetics, soaps, hair oils, agarbatis (incense sticks), scenting clothes and flavouring betel masala, chewing tobacco and bouquets. Kewda water is used for flavouring various foods, confectionaries, sweets, syrups and soft drinks. The oil and attar are considered as stimulant, antispasmodic and administered for headache and rheumatism (Chopra *et al.*, 1956). The kewda is mentioned frequently in Tamil classics as having flowers which neutralize the foul fish odour pervading the sea coast with their strong perfume. Jehangir in his memoirs, *Tuzuk-i-Jahangiri*, described kewda flower for its strong and penetrating scent, which even obscures that of musk (Gandhi, 1989).

Odisha state lying between 17° 52′ N to 22° 45′ N latitude and 81° 50′ E to 87° 50′ E longitude has a prosperous agriculture. Nearly 75 % of its working population are either cultivators or agricultural labourers and remaining 25 % are engaged in the categories of household industry and other works (Anonymous, 1988). The wastelands occupy 13.79 % of the total area of Odisha and are mostly scattered throughout the state. The areas of wastelands in the state mainly accounted for the hilly slopes, degraded forests, over grazed revenue wastelands, oil eroded valleys, drought stricken pastures, saline and marshy lands, etc. The changes in land use leading to creation of wastelands include overgrazing forest degradation and development of marginal cropland due to flood, drought and soil erosion etc. (Fig. 1).

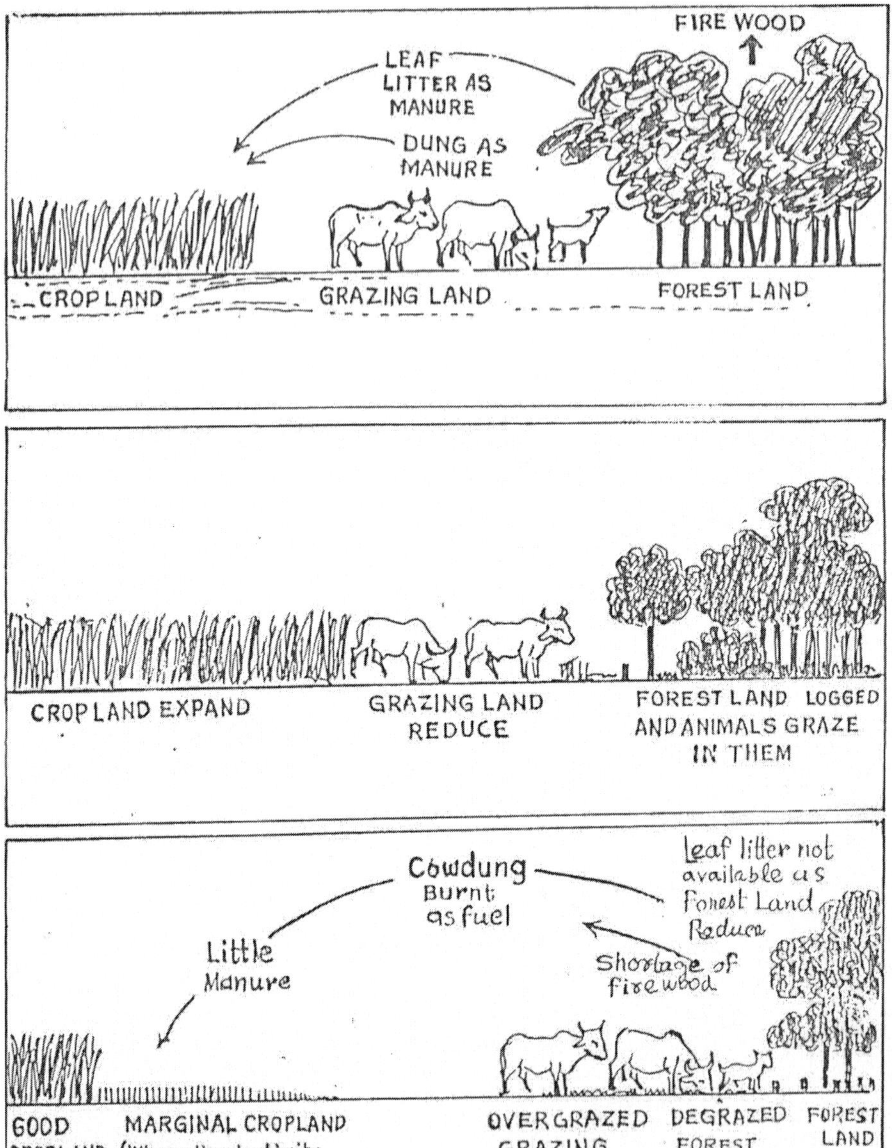

Fig. 1: Changes in land use leading to creation of wastelands

Kewda suited the best for wasteland utilization and development of rural economy of the state as it is commercially viable. It performed well on marginal, waste and saline lands of coastal regions under rainfed conditions, though can also be taken under well drained soil in the interior districts.

Kewda plant is salt tolerant, good soil binder, not browsed by the cattle, once planted can live upto 40 years or more, so a good perennial plant which can successfully be grown on marginal, waste, marsh land and coastal saline sand. They grow wild on riversides or near the sea-shore and need moist saline soil. It is grown along the roadsides, banks of ponds, canals and rivers to check soil erosion and function as fence on the bunds around the crop fields (Plate-1 and 2). Kewda is widespread along the coastal regions of India, but occurs widely in the coastal districts of Odisha, Andhra Pradesh, Tamil Nadu, Gujarat and to some extent in Badayun, Balia and Gajipur districts of Uttar Pradesh. Kewda grows abundantly in the coastal areas of Ganjam district of Odisha in between the rivers Rushikulya on the north and Bahuda on the south. This zone is known as the Kewda belt, which covers an area of about 45 km x 15 km along the coast of Bay of Bengal (Panda *et al.*, 2000). Ganjam, one of the 30 districts of Odisha in the eastern part of India lies between 18° 58′ N to 20° 17′ N and 84° 06′ E to 85° 11′ E. The coastal belt of Ganjam district with an aerial distance of 8-10 km from sea, is about 37,500 ha, out of which Kewda grows luxuriantly on an area of about 4,800 ha (Sahu and Misra, 2007). Kewda flower has high value both in domestic and international markets for preparation of perfumes and medicines. The Agriculture Department of Government of Odisha has identified kewda as a product for export under the aegis of APICOL (Agriculture Promotion and Investment Corporation Ltd.) (Anonymous, 2014).

The commercial exploitation is centred mainly in the Ganjam district of Odisha and the annual turn over of its perfume industry is about Rs. 200 crores accounting for about 90% of India's total production of kewda perfumes and 50% of kewda perfume trade of the world (Dutta *et al.*, 1982) employing over 50,000 farm families with about two lakhs people for collection and distillation of flowers (Subudhi, 2013). Till today, the flowers are mostly collected from natural plantations spreading over 200 villages of 31 gram panchayats in four blocks (viz., Ganjam, Rangeilunda, Chhatrapur and Chikiti) of Ganjam district (Subudhi, 2012).Some initiative may encourage the rural educated youths for establishing the distillation units in the production areas as the cost involvement to establish an indigenous distillery is not very high. Hence, the industry has a great potential for expansion by exploiting the overloaded natural plantations of kewda in the coastal belt.

Plate 1: Thick kewda plantations along roadsides and paddy field bunds in Ganjam district.

Plate 2: Kewda plants planted with *Casuarina* to check soil erosion and as fence to the main crop in Rangeilunda area

Unemployment and under employment are the major causes of rural poverty. Small, marginal and landless families are mainly dependent on agriculture for their livelihood. In drought prone areas, agriculture can provide employment for hardly 100 days in a year and such seasonal empoloyment cannot sustain them throughout the year. Hence, to develop the rural economy, besides kewda perfume industry, small cottage industries can also be set up to utilize the different plant parts of kewda for making various domestic usable items. Hence, the employment it generates in the rural areas and the secondary income it brings to the rural poor can not be ignored. The new areas of natural growing kewda plantations in good concentration should be identified for selection and propagation of superior strains to increase the productivity (Jagadev et al., 1992). Therefore, conscious cultivation of kewda deserves promotion by various agencies, instead of leaving it to grow naturally.

2

Origin and Distribution

Kewda (*Pandanus fascicularis* Lam.) is a native of the old world extending from Africa to Australia (Chittenden, 1956; Lorus and Milne, 1967; Mott, 1975). About 600 species of *Pandanus* are widely distributed in the moist tropics of the old world and kewda is the most widespread species being recorded from Mauritius Islands in the west to Polynesian and Micronesian Islands in the east (Deshaprabhu, 1966).

It occupies an important position in the mythology of Melanesian, Polynesian and Micronesian Islands (Massal and Barrau, 1956). Kewda has been mentioned in the Indian mythology with several available legends. In *Skandha Puran*, it is mentioned that Ketaki (*P. odorattisimus* L.f.) flower is unfit to worship Shiva (Bhattacharjee, 1987). As regards to popularity of Kewda, one of the Gujarati tribal songs says that the sun rises behind the Kewda tree, the moon applies antimony to its eyes, the night is lit with soft moon light. Another mythological legend mentioned that during *Satya Yuga*, there was a fight between Brahma and Vishnu to know, to prove their greatness. Lord Shiva decided a simple method that the one, who can see his head or his feet, shall be greater than the other. Shiva disappeared and became so great that Brahma and Vishnu couldn't see his head or feet and searched in different directions. Vishnu went up to patal, but couldn't find his feet and Brahma had gone up to the highest peak and could not see the head of Shiva. On the returned way, Bramha found a Kewda flower and claimed that he had brought the flower from the head of Lord Shiva. On questioning by Vishnu, Kewda flower also replied that she was adorned by Shiva (Gandhi, 1989).

Jayadev, the famous poet indicated in *Geetgovind* that in the absence of her beloved, the lady's face turns as the flower of Ketaki, that is beautiful, but spinous. Many Sanskrit scholars reported astonishing statements in their poems that why the wasps visit the Kewda flowers, when there is no scope to collect honey. During 16th Century, Bhavaprakash described two types of Ketaki viz., white and yellow (Bhattacharjee, 1987).

Moghul emperor Babar in his *Babar namah* (dated back to 1525) mentioned that musk has the defect of being dried, but the Kewda may be called as moist musk , as it has a very agreeable scent with excellent perfume at the centre of the flower. In India, this flower is so famous for its fragrance that the tree is not planted near a house for fear of attracting snakes (Patnaik, 1993).

Screwpines naturally occur in Madagascar right through the Indian ocean to South-East Asia and Australia and out into the Pacific Islands, where they are commonly seen on the water's edge (Macoboy, 1986) and more than 70 species of *Pandanus* are recorded from this Island (Purseglove, 1975) (Fig. 2) . It is mainly distributed on the sea-coasts and marshy lands of India and the Andaman Island and Malaysia. It also seems to have been widely distributed by human beings within the area and through selection many cultivated forms are reported. About 36 species of *Pandanus* have been recorded in India (Deshaprabhu, 1966) and the most common one, *P. fascicularis* Lam., locally known as kia or kewda, occurs widely in the coastal districts of Odisha, Andhra Pradesh, Tamil Nadu, Gujrat and to some extent in Badayun, Balia and Gajipur districts of Uttar Pradesh (Dutta *et al.*, 1987) (Fig. 3). It is estimated that in the Ganjam district of Odisha, there are about 3 to 4 lakh kewda shrubs (Deshaprabhu, 1966) occupying an area of more than 4800 hectares (Mohammad, 2000) It is also widely distributed in the coastal districts of Odisha (Fig. 4). The plants often form thickets along the roadsides, river banks, canals, bunds of paddy fields and ponds and in coastal sands (Fig. 5).

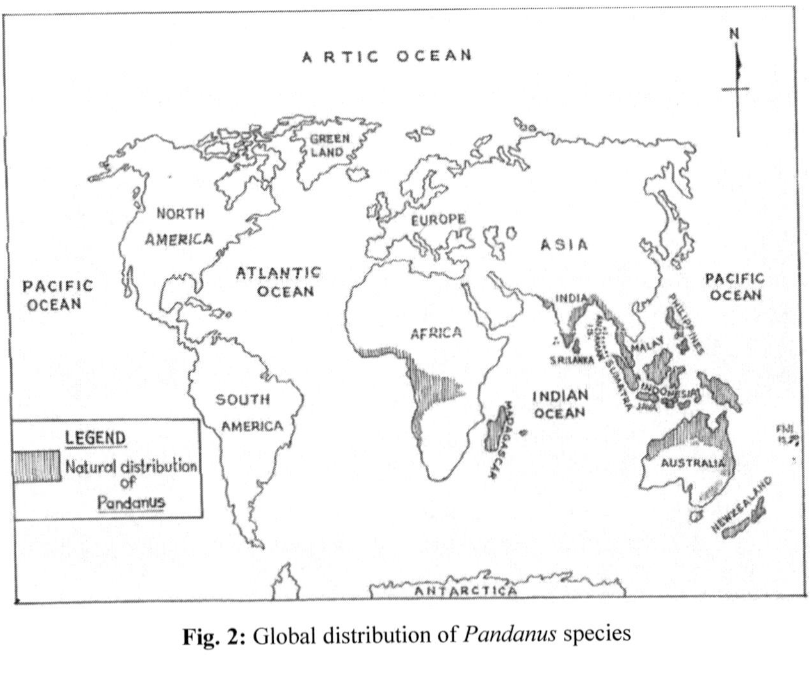

Fig. 2: Global distribution of *Pandanus* species

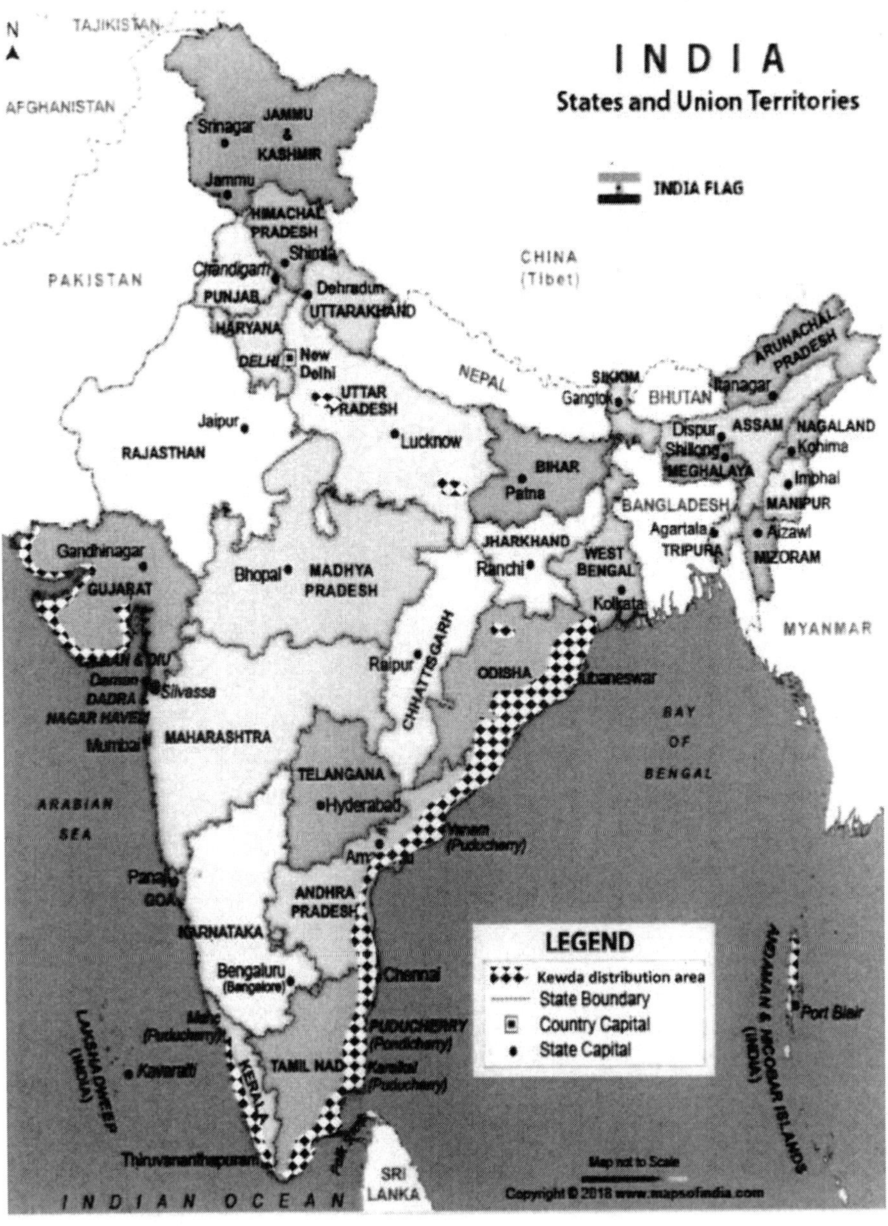

Fig. 3. Distribution of kewda in India

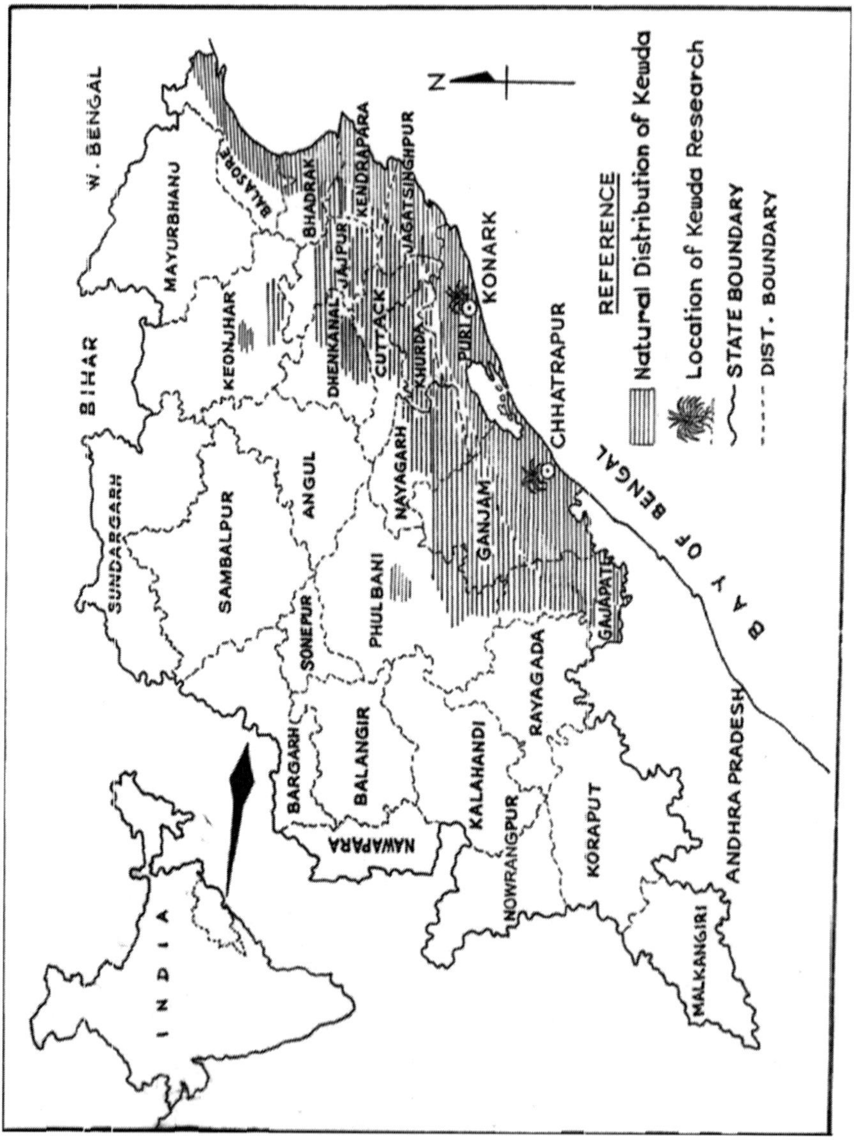

Fig. 4: Distribution of kewda in Odisha (Maharana and Jagadev, 1993)

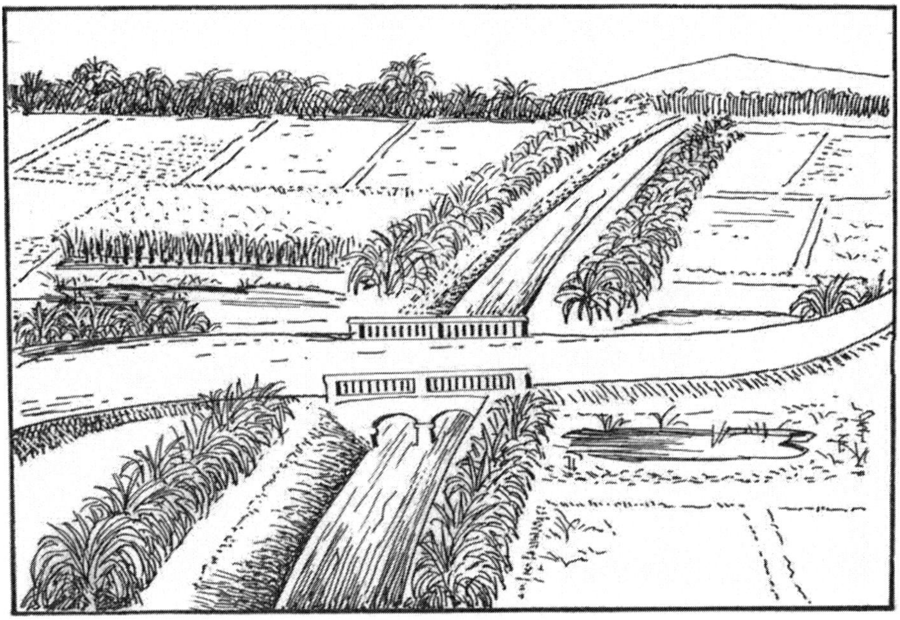

Fig. 5: Kewda plants in natural growth along roadsides, canal and river banks and paddy field bunds

3

Botany and Nomenclature

Botany

Several vernacular names of kewda are reported in different languages all over the world ; viz.

Keora, Ketgi, Keura, Gagandhul in *Hindi*; Keya, Ketaki, Kea, Ketkikea in *Bengali*; Kiya, Kia in Odia; Kewoda in *Gujrati;* Ato, Keto, Covasso in *Konkani*; Keoda, Keora in *Marathi;* Talemara in *Kannada*; Gedaji, Gedangimogali, Gojjangi, Ketaki, Mugali, Mogali in *Telugu*; Kaidai, Talhai, Thasai, Thazh, Kechiya, Kedagi, Kedagai, Muchali in *Tamil;* Kaita, Ketaki, Palliyambu in *Malayalam*; Gandhapuspa, Dalapushpa, Shivadvishta, Ketaki, Suchika pushpa in *Sanskrit*; Leram in *Nicoberese*; Mudukeyiya, wetakey in *Sinhalese;* Keora in *Urdu*; Gulkiri, Kader, Kadi in *Persian*; Satthapu, Tsatthapu in *Burmese*; Nalu in *Nepalese*; Togenashiadan in *Japanese*; Kadar, Kadhi, Kazi in *Arabic* Kaide, Kedige, Ketaki, Gedege, Mundige in *Canarese;* Kedgi in *Deccan;* Vaquois odorant in *French* and Schrauben palme in *German.* (Kirtikar and Basu, 1935; Randhawa and Mukhopadhyay, 1986; Kapoor, 1990).

Systematics

Kewda is a monocot and belongs to family Pandanaceae in the order Pandanales. Of the three families in the order, only the family Pandanaceae includes woody plants and other two families, Typhaceae and Sparganiaceae include monotypic herbaceous marsh-plants with two ranked leaves and creeping stem, are well known in Britain (Rendle, 1979).

Pandanaceae, an old world family includes three genera and about 700 species, largely confined to the tropics. The largest genus, *Pandanus* contains trees and morphological evidences suggest that these plants are more primitive than palms although there is no fossil record (Everett, 1969). The genus has a number of useful species yielding minor products used in tropical Africa, Australia, Islands of the Indian ocean and the Pacific. The genus *Freycinetia* has about 100 species extending from Sri Lanka through South-Eastern Asia to Polynesia and

New Zealand (Plate 3 to 7), are mostly climbing shrubs with flowers and inflorescence like *Pandanus,* but the fruit is a berry, not a drupe. The genus, *Sararanga,* with only two species in the Philippines, New Guinea and Solomon Islands, lacks stilt roots and its inflorescence is a panicle with pedicelled flowers and fruit is a drupe (Willis, 1960; Purseglove, 1975).

Plate 3: Freycinetia arborea (Fiji) (Graf, 1986)

Plate 4: Freycinetia funicularis male inflorescence (Java, Indonesia) (Graf, 1986)

Plate 5: Freycinetia multiflora (Phillipines) Climbing pandanus (Graf, 1986)

Plate 6: Freycinetia rigidifolia (Thailand, Malaya, Borneo) (Graf, 1986)

Plate 7: Freycinetia stenophylla (New Guinea) (Graf, 1986)

Cytology

Purseglove (1975) reported the basic chromosome number for the genus *Pandanus* as 30 and 32.

Nomenclature

The genus, *Pandanus,* popularly known as either screwpine or walking palm is neither a pine nor a palm, but is once thought to be related to the pine apple (Macoboy, 1986). The common name, screwpine, is due to the saw-edged leaves borne in a 3-rowed phyllotaxy like a screw up the stem that are usually twisted, appearing the leaves to run in well marked spirals towards the tips of the branches, which together with the large cone –like fruits give a vague resemblance to pineapple (Mott, 1975; Purseglove, 1975; Heywood, 1978).

The genus, *Pandanus,* is a Latinized form of a Malayan word *Pandan,* means stick man (Gandhi, 1989). The name *Pandanus* is derived from *Pandang,* a Malayan word signifying conspicuous (Chittenden, 1956; Macoboy, 1986). *Pandanus fascicularis* Lam., the most common of the 36 species found in India (Deshaprabhu, 1966) is described under the name *P. tectorius* Soland. in most Indian floras. Stone (1976) pointed out that *P. tectorius* is a Polynesian plant (Plate-8), specifically distinct from the Indian plant by the smaller green leaf margin teeth and lack of conspicuous fleshy shoulders that characterise the fully ripe pistillate phallanges of *P. fascicularis.*

Position of Species

Kewda (Plate 9) is widely distributed and often cultivated commercially in the Ganjam district of Odisha for production of flowers which are extensively used in the perfume industries. Other species like *P. odoratissimus* L.f., known as 'Ketaki' (Plate 10) and *P. latifolius* Perr. (Syn. *P. amaryllifolius*), locally known as 'Annapurna' (Plate 11) are also found in the state. In the whole genus *Pandanus*, *P. amaryllifolius* Roxb. is the only one with scented leaves (Stone, 1978). Annapurna is used for flavouring the cooked rice as its leaves are highly aromatic like basmati rice and spineless.

The major volatile compound 2-acetyl-1-Pyrolline in the leaves has commercial significance as a major flavour constituent , which is similar to the chemical responsible for characteristics aroma in basmati rice also (Ghosh, 2001). The leaves are also rich in magnesium and iron. There is a great scope to explore Annapurna leaves to be used as a convenient substitute of expensive basmati aroma in non-aromatic rice (Rayaguru *et al.*, 2008).

Plate 8: Pandanus tectorius (New Guinea) (Graf, 1986)

Plate 9: Kewda (*P. fascicularis*) Plant

Plate 10: Ketaki (*P. odoratissimus*) Plant

Plate 11: Annapurna (*P. latifolius*) Plant

Various literature mentioned that *P.odoratissimus* L.f., locally known as Ketaki is synonymous to *P. fascicularis* Lam. (Kewda) nomenclaturally (Dutta *et al.*, 1987); but it is revealed that these two are distinct species as regard to their morphology and adaptation. The distribution of 'Ketaki' is limited to the deep fertile soil near perennial water flowing nalla site of inland districts, while 'Kewda' is commonly grown under various situations. According to observations, Ketaki is possibly a superior selection of male plant of Kewda.

A comparative study on variations in these two aromatic plant species showed that ketaki flowers (Plate-12) are golden yellow and strongly scented, while kewda flowers (Plate-13) are cream coloured and sweetly scented (Table 1).

Morphology

The plant comprises a bushy shrub or growing into a small tree, erect or rarely prostrate, simple or 2-3 chotomously, branched, trunk annular, cylindric, 3-6 m high, supported by aerial stilt roots, having conspicuous membranous root-caps (Chittenden, 1956; Willis, 1960); leaves : linear and acute, 0.9-1.6 m long, 4-5 cm wide, coriaceous, sessile, sheathing base, tristichous, edges and mid-rib usually spinous except in thornless types, spines on margins point forward and on mid-rib usually retrose; *inflorescence* axillary terminal, compound, racemose (Rendle, 1979); flowers: densely clustered, dioecious, naked, sessile in compact heads, without intermixed bracts, enclosed in caducous spathes, densely crowded on spikes; perianth none; male spadices very sweet scented, 25-40 cm long, sub-sessile cylindric spikes 5-10 by 2.5-3.8 cm enclosed in long pale yellow or white caudate, acuminate, fragrant spathes, spicately arranged on the floral axis, stamens numerous, filaments: free or connate, anthers erect, basifixed, pistillode small or obsolete; female flowers : always terminal, solitary, 5 cm diameter, stout, unbranched, subtended by 3 fariously arranged white spathes, carpels numerous, syncarpous, staminodes small or none ovary of one to a variable number of carpels with corresponding number of cells, ovule solitary in each cell, ascending from the base of a parietal placenta, stigmas short, subsessile, papillase, reniform and yellow, fruit a pendant pineapple, inflorescence usually develops to a massive syncarp, of one to several celled woody or fleshy drupes angled by pressure with neighbouring drupes, drupes deciduous, singly or in masses from a fleshy receptacle, upper half of each carpel or drupe is hollow or filled with a spongy pith like tissue, apex of each drupe or carpel distinct pyramidal, conical, crowned by hardened style or stigma, whole forming a large globose, ovoid and clavate syncarp, 15-25 cm long and 10-20 cm diameter, deep orange coloured when ripe on peduncle of 12.5-15.0 cm long; seed albuminous, minute, testa striated, albumen hard fleshy, pericarp rich in fibre, embryo minute.

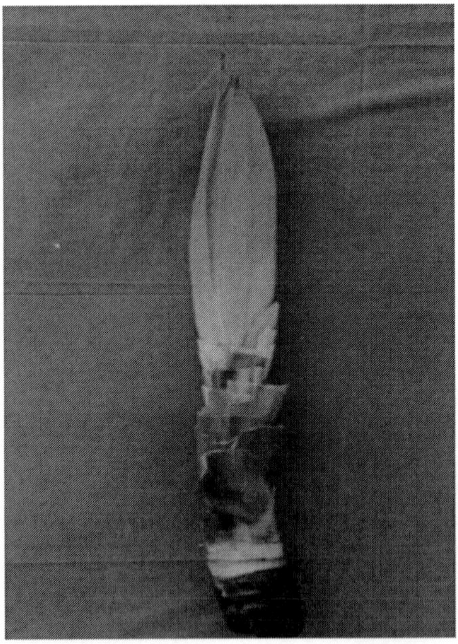

Plate 12: *Ketaki (P. odoratissimus)* flower

Plate 13: *Kewda (P. fascicularis)* Male flower

Table 1: Comparative study between kewda (*P. fascicularis* Lam.) and ketaki (*P. odoratissimus* L.f.)

Character	Kewda	Ketaki
Leaf length (cm)	70.0	81.0
Leaf width (cm)	3.5	3.4
Leaf thickness (mm)	0.5	0.8
Leaf weight (g)	7.8	8.5
No.of thorns/10 cm leaf margin	8.0	6.0
Angle of projection of thorn on leaf margin	45°	75°
Time taken for discolouration of leaf in water (days)	7.0	8.0
Time taken for browning of leaf at room temp. (days)	4.0	5.0
Leaf colour	Green	Green
Leaf stiffness	Less	More
Flower length (cm)	48.0	65.0
Flower weight (g)	145.0	250.0
No.of bracts/flower	16.0	16.0
Flower colour	Cream	Golden yellow
Flower aroma	Sweetly scented	Strongly scented

4

Species and Varieties

The screwpines are tropical shrubby plants grown for the highly scented flowers or the ornamental foliage. The ornamental foliage types are generally cultivated in the green house and protected from direct sun. The variegated types loose the colour in a rich soil and hence should be grown in a rather poor soil containing just sufficient nutrients and moisture. Some species are good for potting and indoor decoration. The common screwpine is used as a hedge also (Randhawa and Mukhopadhyay, 1986)

The screwpine (*Pandanus)*, is highly polymorphic with numerous forms and varieties and selected in various countries for specific uses as maintained by vegetative propagation, e.g. *P. leram* (Plate 14), the Nicober-bread fruit, is grown in the Andaman Islands for its edible fruit providing a starchy food that is enjoyed locally; another species, *P. utilis* (Plate 15), on the remote Island of Mauritius, provides fibre from the leaves, used in making bags and a species from Polynesia, *P.veitchii* (Plate 16) is most commonly grown in green houses and outdoors as an ornamental, where the winter weather is mild and in the species the prop roots and branching stems are particularly well developed attracting attention from many people who might not otherwise notice the stranger from the tropics (Lorus and Milne, 1967). However, these three species give a striking foliage effect and especially useful in seaside gardens as *P. utilis* with green leaves; *P. veitchii* with banded white and *P. leram* with even longer leaves (Thomas, 1965). An unarmed form, forma *laevis* warb., is cultivated for the fragrant bracts of the inflorescence, while forma *samak,* a group of prickly *Pandanus,* is preferred for the tough leaves suitable for matting; another form with longitudinal yellow bands on leaves (forma *variegatus)* is grown for ornamental purposes (Deshaprabhu, 1966; Ambasta, 1986). In the Pacific Islands, some of the forms (e.g *pulposus)* are cultivated for their edible fruits (Purseglove, 1975; Ambasta, 1986). Burkill (1966) states that the common wild armed form of the sea-coasts was var. *littoralis* and that flower growers had selected various races for cultivation, which included the prickly var. *samak* and thronless var. *laevis,* grown in India, Malay and Indonesia, are used for making mats.

Plate 14: *Pandanus leram (Nicober breadfruit)* (Graf, 1986)

Plate 15: *Pandanus utills (Madagascar)* (Graf, 1986)

Plate 16: *Pandanus veitchii (Polynesia)* (Graf, 1986)

The number of species of the genus *Pandanus* in commercial cultivation are very few, although many names occur in horticultural literature. In South India, *Pandanus* is known so far to have six wild or naturalized species, viz., *P. fascicularis* Lam., *P. foetidus* Roxb.,*P. furcatus* Roxb., *P. kaida* Kurz., *P. thwaitesii* Mart. and *P. unipapillatus* Dennst. (*P. canaranus* Warb.) (Bhat, 1992). Without flowers and fruits, it is difficult to know what species are actually in cultivation, or how accurate may be the popular descriptions and illustrations. For the same reason it is impossible to construct and accurate botanical key that will be of practical use to the gardener. Some of the good garden kinds are unplaced botanically, particularly the variegated or striped leaved kinds, which are sterile or the fructification insufficiently studied. However, the species which occur in various parts of the tropics and subtropics include the following species (Bailey, 1950; Chittenden, 1956; Deshaprabhu, 1966; Purseglove, 1975; Tanaka, 1976; Wickham, 1982; Randhawa and Mukhopadhyay, 1986; Southwell and Harris, 1992).

(1) P. andamanensium Kurz.

Vernacular : Hindi –Keora

An evergreen tree, 9-15 m tall, common in the low moist swampy areas in the Andaman Islands. A fibre is made from it by the Andamanese for making various articles of apparel, such as the 'tails' worn by the women and the fruit is eaten after cooking.

(2) P. aquaticus F.V. Muell.

Roasted seeds are eaten by the aborigines of Queensland, occurs in Australia.

(3) *P. artocarpus* Griff

Grows to a height of 20 m, occurs commonly in Southern Malay and neighbouring parts of Indonesia. Large, very tough leaves, about 6 m x 15 cm, used for thatching and making mats for covering carts, cut into finer ribbons and plated into hats.

(4) *P. baptistii* Hort.

A leafy short-stemmed plant, leaves narrow, spirally arranged, arching beautifully and tapering to a long point, leaves are nearly 2.5 cm wide, curved, unarmed, long acuminate, blue-green in colour with 1-4 longitudinal yellowish white stripes, having no spines, common in New Britain Islands near New Guinea (Plate 17).

(5) *P. barklyi*

Slender tree, 5-8 feet height, leaves 1-2½ feet long, usually about 2" wide, margins and mid-rib with red spines, fruit of 70-100 drupes, purple, 4-6 seeded, common in Mauritius.

(6) *P. boninesis* Wars.

Fruit is chewed, seeds eaten roasted like pine seed or used as a source of edible oil, found in Micronesian Islands.

(7) *P. brosimus* Merr. & Perry.

Cultivated in the highlands of New Guinea, grown for seeds which have a pleasant flavour and rich in oil, tall erect tree with stout trunk to 30 m height gigantic stilt roots, broad-leathery, saw edged leaves 5m long with drooping tips, pineapple like fruit as a source of food to the mountain people of New Guinea and highland Papuans (Plate 18).

Plate 17: *Pandanus baptistii (Melanesia)* (Graf, 1986)

Plate 18: *Pandanus brosimus (New Guinea)* (Graf, 1986)

(8) *P. bullii*

Small tree, leaves 2 ft. long, arching, dark green, margin and nerves above weakly spiny, midrib smooth beneath, common in New Caledonia.

(9) *P. candelabrum* Beauv

A swamp tree to 10 m along West African coast, leaves used for making mats, baskets and bags.

(10) *P. caricosus* Kurz.

Small tree, short height, charming fountain like slim leaves, leaves 5-8 ft by about 2 in., glaucous beneath; margins and dorsals costae densely spinulose serrate, spines very small and acute, white flowers common in Java.

(11) *P. ceramensis* – A synonym of *P. bullii*

(12) *P. ceylanicus*

Similar to *P. furcatus,* but stem more slender, branches more erect, foliage narrower, common in Sri Lanka.

(13) *P. conoideus* Lam.

A low land staminate plant. Pericarp of fruit is dark in colour and contain fat which is a source of vitamin A. Fruit is cooked and used like condiment, occur in Malay and Indonesia. The plant appears to grow wild, but is reported to be cultivated in some areas.

(14) *P. decorus*

Differs from *P. bullii* by its lighter green leaves, spiny on mid rib beneath. Common in New Caledonia.

(15) *P. dubius* Spreng.

Knob-fruited screwpine, kernels are sometimes eaten, occur in Philippines, Indonesia and Micronesia.

(16) *P. duriocarpus* Martelli.

Fruit is edible, occur in Micronesia.

(17) *P. dyerianus* – A synonym of *P. baptistii.*

(18) *P. edulis* Thouars.

Fruit is edible, occur in Madagascar.

(19) *P. elegantissimus* – A synonym of *P. utilis.*

(20) *P. foetidus* Roxb.

Vernacular : Hindi -Keor-kanta; Bengali – Keya-kanta, kotki-kanta.

A prostrate shrub 0.9-1.5 m tall, glaucous green leaves with short and sharp spines on margin, occurs in West Bengal, Assam and Khasi hills of India, common in tidal forests and wet places, frequently planted near field as a hedge plant in Srilanka,, the spathes are yellow and the flowers are foul smelling.

(21) *P. forsteri* Moore and Muell.

(Syn. *P. fosterianus* Hort.; *P. moorei F. Muell.)*

Habit of *P. tectorius,* 30 ft. tall, branched at top, with aerial roots, leaves 3 ft. by 2 in or more, short appressed-spinulose, occurs in lord Howe's Islands.

(22) *P. furcatus* Roxb.

(Syn. *P. horridus* Blume; *P. spinifructus* Dennst.; *P. urophyllus* Hance.)

Vernacular: *Marathi* –Ran-keura

An ornamental tree, 3-9 m tall, leaves dark green, with short, stout sharp spines on the margins and mid rib spathes golden yellow, common in Sikkim, Himalayas, Assam, Khasi and Naga hills and Konkan and Burma. Plants form impenetrable thickets in marshy places near water courses. In Burma, leaves used for making mats, young leaves as an antidote for poisoning, wood employed to make floats for fishing nets.

(23) *P. graminifolius*

An attractive species, which is sometimes confused with *P. pygmaeus.* It is about 0.6 m height, leaves 30-45 cm long, 1.25 cm wide, greyish green colour, slightly spiny on margins and mid rib beneath.

(24) *P. heterocarpus* Balf. f. (Syn. *P. ornatus* Bull.)

Branched tree with slender trunk, 6 m tall, very numerous roots and spreading branches, leaves lanceolate, acuminate, dilated and clasping at base, erect-spreading, flat margin covered with small red, slightly incurred spines, occurs in Mascarene Islands.

(25) *P. houlletii* Carr.

A forest species from Southern Malay, soft sweet fruits flavoured like pineapples and edible.

(26) *P. julianetii* Mart. (Hyndmans, 1984)

Grown in highlands of New Guinea, has edible seeds having pleasant flavour and rich in oil like *P. brosimos.* Plant may continue bearing for 60 years. Seeds yield edible oil.

(27) *P. kafu* Martelli (Syn. *P. fragrans)*

Fragrant-fruited screwpine, kernel is eaten, occur in Micronesia.

(28) *P. kirkii* Rendle

Branched tree, 10 m tall, common on coasts of East Africa and Zanziber, soft durable leaf strips used for making conical hats worn by local fishermen, fragrant male inflorescences much sought after local women.

(29) *P. labyrinthicus*

About 3.6-5.4 m height, spreading branches and aerial roots form a very intricate system, leaves at ends of branches 11-15 cm long, 1" wide, very acute, glaucous beneath, densely white spiny on margins and midrib beneath, fruit red-orange when ripe, elliptic, about 10 cm long , Common in Sumatra.

(30) *P. latifolius* Perr. (Syn. *P. amaryllifolius)*

Vernacular name – Pandan wangi (Malaya, Java) (Stone, 1978) and Annapurna in Odia. A small tree, introduced from Malaysia and now grown on west coast of India, used for flavouring cooked rice (Singh *et al.,* 1983). It is also known in West Indonesia, Philippines and Srilanka to flavour rice and jellies in cooking (Stone, 1978). But the scent in Pandan wangi leaves is apparently not an essential oil. De Jong attempted to distill the essence from 70 kg Pandan leaves but with no success (Teng *et al.,* 1979)

(31) *P. leram* Jones Vernacular : Nicober bread fruit.

A slender tree, 12-15 m tall, common in Andaman and Nicober Islands, lower fleshy portion of globular fruit weighing 13-18 kg is edible and used for making flour for bread by the local people, by boiling in water and pressing, fruits produce a mealy mass strongly resembling the flavour of apple marmalade.

(32) *P. lindenii* (Syn. *P. ornatus)*

Elegant small plant, leaves form a dense crown, erect then drooping at ends, 0.9-1.4 m long, shining green with pale, narrow, translucent, longitudinal lines, brown at filiform tip, spines short and slender, white on margin, black on midrib beneath. Common in Mascarene Islands.

(33) *P. mellori* Boden – Kloss. (Syn. *P. amaryllifolius)*

The fruit makes a bread, occurs in Andaman and Nicobar Islands.

(34) *P. obliquus* Kanehira

Fruit is eaten raw, cooked or preserved, occurs in Marshall and Gilbert Islands.

(35) *P. odoratissimus* Linn.f.

(Syn. *P. tectorius* soland. Ex. Parkinson; *P. fascicularis* Lam.)

The common screwpine, most important widespread species, which is often cultivated commercially and occurs widely in Madagascar and in coastal regions of India, Arab, Iran and Burma. It is a shrub with fragrant flowers and perfumes obtained from its male spadices. The scent of the flower, it is alleged, attracts cobras, which remain hidden in the flowers and hence one should handle the flowers with proper inspection.

(36) *P. odorus* Ridl.

A small screwpine seldom more than 15 m tall, commonly cultivated by Malays for its fragrant leaves, freshly chopped leaves used in cooking and has a number of local medicinal and magical uses.

(37) *P. pacificus* Hort.

Leaves broad, dark green, shining, abruptly narrowed to a tail like apex, small fine spines on margins, common in Pacific Islands.

(38) *P. patina* Martelli

Seeds are eaten by cracking, occurs in Micronesia.

(39) *P. pedunculatus* R. Br.

Bread fruit, fruit is eaten, occurs in New Guinea and Australia.

(40) *P. polycephalus* Lam.

Young leaves are eaten raw by the natives, occurs in Malay.

(41) *P. pristis*

Clustering, rosette 2-4 m height, dark glossy leathery foliage with prominent pale spines, leaves spreading out horizontally and recurving beyond the middle, begin to flower when 1m of height, occurs in Madagascar (Plate 19).

Plate 19: *Pandanus pristis (Madagascar)* (Graf, 1986)

(42) *P. pulposus* Martelli

Fruit is eaten raw, cooked and also preserved, occurs in Pacific Islands.

(43) *P. pygmaeus* Thouars.

Low spreading shrub, about 0.6 m in height, but sending out from the base numerous horizontal, rooting annulated branches, leaves about 45 cm long, 1.2 cm wide, spirally arranged in 3's, linear-subulate, with a clasping base; margins and keels fringed with small fuscous spines, occurs in Madagascar. Sometimes it is known by *P. graminifolius* name in gardens.

(44) *P. reflexus* C. Koch.

Tree, about 3.3 m branched, trunk 0.9-1.8 m tall, leaves reflexed, acute 5-6 ft, long, dark green shining, spines long, whitish, those on the midrib of the lower side reversed, occurs in East African islands.

(45) *P. rockii*

Tall, freely branching, forming stilt roots with smooth branches topped by rosettes of leathery lanceolate leaves, margins finely serrate, occurs in Hawaii (Plate-20).

(46) *P. sanderi Roehrsianus*

The leaves are sword-like, armed with fine spines, 1.2 m long, shining green with length-wise stripes and bands of light and golden yellow, the young foliage is golden orange in colour, occurs in Malayasian archipelago (Plate-21).

Plate 20: *Pandanus rockii (Manoa Valley)* (Graf, 1986)

Plate 21: *Pandanus sanderi Roehrsianus (Malayasia)* (Graf, 1986)

(47) *P. sinensis* Warb.

Nuts are chewed by the natives, occur in China and Micronesia.

(48) *P. tectorius soland*

Slender, branched, 6 m tall with flexuous trunk supported by brace-roots, light green fibrous leaves, pendulous clusters of seed, occurs southern tropical Asia to Polynesia.

(49) *P. thwaitsii* Mart.

A common under shrub with thin, densely matted stems, flexuose male spadix found in rain forests on the Siddapore ghats of N. Kanara, at Bantwala inS. Kanara upto 300 m and in Travancore, fragrant white flowers, fibre from leaves used for nets, brushes and fruit pulp is edible.

(50) *P. utilis* Bory.

(Syn. *P. distichus* Hort.; *P. elegantissimus; P. flabelliformis Carr.; P. mauritianus* Hort.; *P. sativus* Thouars.)

A stout branching tree, 15 m tall, native of Madagascar and introduced elesewhere as an ornamental, in Indian gardens grown as a decorative plant, best suited by its size to large offices, leaves are strap-like, leathery, long and spirally arranged, spiny at the margins, about 2 m long, used for thatching and making baskets, mats, hats and other domestic articles, roots used for veneral diseases, fruits produce a starchy food, palatable when cooked, commonly cultivated in Mauritius, where it is used for making sugar bags.

(51) *P. utiyamai* Kanehira.

Fruit is eaten raw or cooked, cooked one is made into preserves, occur in Carolin Islands and Gilbert Islands (Micronesia).

(52) *P. vandermeeschii* Balf. f.

A much branched tree, 6.5 m tall, leaves stiff, sub-erect, 1 m long, 5 cm wide, very glaucous, margins red and thickened, spines strong, red, midrib red, prominent, spiny, occurs in Mauritius.

(53) *P. variegatus* Miq.

Stout branched, emitting roots, leaves narrow, lanceolate, pale green, striped or variegated with white or pale green or the younger ones all white, prickles whitish and purplish, occurs in Polynesia.

(54) *P. veitchii* Hort.

Plants variegated, young plants popular for foliage in home decoration, leaves 0.6 m long, broader than *P. utilis,* somewhat recurved, spiny-toothed, dark green in centre, margined with broad bands of white or silvery white, apex long-acuminate, occurs in Polynesia.

(55) *P. zeylanicus* – A synonym of *P. ceylanicus*

5

Breeding and Crop Improvement

No breeding work has so far been done on kewda except selection of some promising morphotypes. Varietal improvement in kewda by breeding poses problems mainly due to dioecious nature and long life cycle of the plants, besides heterozygosity. Lack of promising strains is a major drawback. The wild growing population of kewda has a measure of variation and this has never been documented and utilized, which naturally offer a good scope for improvement of kewda by resorting to selection and subsequently propagating them vegetatively. Since a wide range of variation is available and the plants are mostly propagated by vegetative means, it is worthwhile to collect genetic stocks, access variability in morphological characters and attempt a breeding programme after selecting the desirable types for high flower production as the ultimate economate economic value of the plantations depends on the quantity of flower production to be used industrially in the distillation units.

Varieties

There is no distinct named varieties available in kewda. Although variations are more spectacular in kewda (St. John, 1965), a detailed study of these variants and their systematic position is still lacking. To study the presence of variability followed by selection of types in the natural strands of kewda plantations under farmers' field condition, a detailed survey was made in the coastal districts of Odisha (Table 2). Based on the variability present, six distinct morphotypes were selected (Jagadev *et al.,* 1993), which exhibit variations in different quantitative and qualitative characters (Table 3), particularly flower bearing, weight, colour and aroma of flower and interestingly, in adaptability with regard to location, land type and soil texture. The selection of clones was done on the basis of leaf pigmentation (dark green/light green) and presence of thorn or absence (thornless) and size of thorns (large/small) on leaf margin. The thornless types were identified in the villages namely Markandi and Gunupur region (Ganjam district). The mala or dwarf types were short in height, bushy and spreading in growth habit, narrow leaved, occasionally flower only during rainy season when plants grow older. Flower production, the most important economic trait, was maximum in darkgreen types and minimum in thornless and mala (dwarf) types, while the flower was strongly scented in dark green large thorny type. In adaptability, large thorny types prefer sandy soils closeness to sea, while small thorny types perform well under saline situation and thornless types under upland condition at nalla site (flowing water side) and mala types under inland situation. Deshaprabhu (1966) reported that fertile, well-drained soil was preferred for raising scented types in India. Hence, a comprehensive survey is desirable to identify some promising genotypes from natural variability existing along the coastal belt, which possesses the largest acreage under kewda plantations for production of kewda flowers.

Survey and Selection of Types

To study the presence of variability with a view to locate some promising types. From the natural plantations of kewda under farmers' field condition, the villages in different regions of the coastal districts of Odisha were surveyed (Table 2)

During the survey, based on the variabilities present, six different morphotypes, viz. dark green large thorny (Plate 22), light green large thorny (Plate 23), dark green small thorny (Plate 24), light green small thorny, thornless (Plate 25) and mala (dwarf) (Plate 26) types were selected, which exhibit distinct variations in 19 important characters (Table 3), particularly flower bearing, weight, colour and aroma of flower and interestingly, in adaptability with regard to location, land type and soil texture. Flower production, the most important economic trait, is maximum in dark green small thorny type followed by dark green large thorny type and very poor in mala type, while aroma of flower was strongly scented in dark green large thorny type. In adaptability, large thorny types perfer sandy soils closed to sea, whereas small thorny types perform well under saline situation and thornless types under upland condition at nalla site (flowing water side) and mala types under inland situation. Deshaprabhu (1966) reported that fertile and well drained soil was preferred for raising of scented types. Hence, a comprehensive survey is thought desirable to identify some promising genotypes amongst the naturally variability existing along the coastal belt of Odisha, which possess largest acreage under kewda plantations for production of flower and kewda perfume. The breeding and selection strategy for crop improvement can also obviously be habitat and genotype specific.

Plate 22: A dark green large thorny type kewda plant

Plate 23: A light green large thorny type kewda plant

Plate 24: A dark green small thorny type kewda plant

Plate 25: A thornless type kewda plant

Plate 26: A mala type kewda plant

Table 2: Survey of kewda plantations in different coastal districts of Odisha

Name of the district	Name of the region	Name of the village surveyed	Remarks, if any
1. Ganjam	i) Chhatrapur	Arjupalli, Chhatrapur, Agraharam, Samakhandi, Purunabandha, Puintala,Badaputi, Basanaputi, Uplapati, Palur, Kanakana and Prayagi.	-
	ii) Gopalpur	Rangeilunda, Golabandha, Gopalpur, Korapalli, Mandiapalli, Haripur and Badapur.	-
	iii) Tulu	Keluapalli, Tulu, Panapalli, Indrakhi, Gunupur, Markondi, Kirtipur, Loudigaon, Dumudumi, Solabindha, Hantulu, Reddytulu, Phasenapara, Dhepanapara, Mandrajpur, Eksingh and Dankalpadu.	Less thorny and thornless types are identified in Indrakhi and Gunupur villages respectively. These types produce very few flowers as compared to thorny types.
2. Puri	i) Konark	Bhanpur, Nuadhi, Siripur, Kurujang, Kaimati, Kunjamatha, Santiamath, Raulapatna and Chandrabhaga.	-
	ii) Puri	Pipili, Dandamukundapur, Satsankha, Patnaikia, Brahmagiri, Sakhigopal, Nimapara and Gop	-
3. Khurda	i) Bhubaneswar	Jagamara, Aaiginia, Patrapada, Tomando, Madanpur, Dumuduma, Kaushalyaganga and Uttara.	-
	ii) Chilika	Dahikhia, Gambhari, Soron, Karadi, Kalupada, Injanpur and Banapur.	-
4. Cuttack	i) Phulnakhara	Phulnakhara, Trinath bazar, Niali, Hansapal and Sankhatras	-
	ii) Balikuda	Balikuda, Naugaon and Kaniha	-
	iii) Choudwar	Choudwar, Biswanahakani, Badachana and Chhatia	-
5. Jagatsinghpur	i) Paradeep	Paradeep, Rangiagarh and Kujang.	

Table 3: Variability in flower production and its ancillary characters of kewda

Character	Character values in different types.					
	Dark green large thorn	Light green large thorn	Dark green small thorn	Light green small thorn	Thornless	Mala type
Length of leaf (cm)	70.0	64.5	65.0	68.0	75.0	55.0
Breadth of leaf (cm)	3.5	3.8	4.0	4.0	3.0	2.5
Thickness of leaf (mm)	0.5	0.6	0.4	0.6	0.4	0.8
Weight of leaf (g)	8.0	7.5	7.5	7.3	9.5	5.5
No. of thrones/10 cm leaf margin	9.5	10.0	15.0	13.0	-	10.0
Angle of thorn projection	45^0	45^0	85^0	85^0	-	50^0
Time taken for discolouration of leaf in water (day)	7.0	6.0	7.5	6.0	8.0	6.0
Flower production ability in 25 years old plant / year	25.0	20.0	28.0	24.0	10.0	3.0
Length of flower (cm)	45.0	55.0	40.0	55.0	65.0	42.5
Weight of flower (g)	85.0	92.5	50.0	110.0	175.0	135.0
No. of androecious fesicles/flower	16.0	13.0	13.5	14.0	14.0	11.0
Length of androecious fesicle (cm)	6.5	6.0	6.0	6.5	7.0	5.0
No. of bracts / flower	19.0	18.0	14.5	18.0	17.0	13.0
Colour of leaf	Dark green	Light green	Dark green	Light green	Pale green	Light green
Stiffness of leaf	Less	Less	Stiff	Less	Tender	Tender
Colour of thorn	Red	White	Red	White	-	White
Colour of flower	Yellowish white	White	Yellowish white	White	White	White
Aroma of flower	Strongly Scented	Sweet Scented	Sweet scented	Sweet scented	Less scented	Less scented
Adaptation	Sandy and Sandy loam soil close to sea	Sandy and Sandy loam soil close to sea	saline situation	Saline situation	Uplandsituation at nalla (flowing water) site	Inland situation

Variability and Character Association Studies

Kewda plant normally takes more than five years after planting for initiation of flowering (Purseglove, 1975). The relationship of flower production to its attributing characters is of great importance for effective selections and it will also help for development of appropriate cultural practices for boosting high flower production.

Studies on variation and genetic parameters revealed that high estimates of genotypic coefficients of variation, heritability and genetic advance are observed for leaf width, number of thorns on leaf margin and flower weight (Table 4) indicating their reliability for effective selections for high flower production. Flower production per plant is found to be significantly positively associated with leaf width and negatively with flower weight both at genotypic and phenotypic levels (Table 5, Fig. 6 and 7). Leaf width and number of thorns are the best contributors to flower production (Table 6 and Fig. 8). Regression studies revealed that an increase of one cm in leaf width increased flower production by 15.09 per plant, while an increase of 1g in flower weight and 1 mm in leaf thickness decreased the flower production by 0.18 and 34.78 per plant, respectively (Table 7) (Jagadev *et al.,* 1993).

Table 4: Estimates of genetic parameters in kewda

Character	Mean ± SEm	Rangw	Mean square	PCV (%)	GCV (%)	H (%)	GA (% of mean)
Leaf length (cm)	66.25±3.76	51.4-78.1	89.95*	10.89	9.29	72.89	16.35
Leaf width (cm)	3.47±0.17	2.4-4.1	0.73**	17.79	17.10	92.39	33.85
Leaf thickness (mm)	0.55±0.13	0.3-0.9	0.05	29.34	20.41	48.39	29.25
No. of thorns/10 cm of leaf margin	9.58±1.92	0.0-17.0	53.28**	55.71	51.98	87.07	99.92
Flower weight (g)	110.75±12.45	47.0-181.0	3578.75**	39.01	37.36	91.70	73.70
Flower length (cm)	51.00±4.95	34.0-67.0	196.20*	20.60	18.17	77.77	33.00
No. of bracts/flower	16.58±1.77	12.0-21.0	10.88	15.98	11.86	55.11	18.14
Flower production/plant	18.33±3.51	2.0-31.0	190.93**	55.00	51.55	87.86	99.55

*,** Significant at 5 % and 1 % level, respectively

Table 5: Genotypic (G) and phenotypic (P) correlation coefficients among various traits in kewda

Character		Leaf width	Leaf thickness	Thorns/10 cm leaf margin	Flower weight	Flower length	Flower production/plant
Leaf length	G	0.322	- 0.795	- 0.507	0.191	0.727	0.385
	P	0.329	- 0.947**	- 0.535	0.164	0.553	0.349
Leaf width	G		- 0.495	0.630	- 0.748	0.016	0.961**
	P		- 0.405	0.518	- 0.694	- 0.023	0.913**
Leaf thickness	G			0.156	0.301	- 0.352	- 0.674
	P			0.293	0.145	- 0.151	- 0.462
No. of thorns/10 cm of leaf margin	G				0.854*	-0.789	0.607
	P				-0.804	- 0.656	0.549
Flower weight	G					0.693	- 0.819*
	P					0.585	- 0.818*
Flower length	G						-0.177
	P						-0.218

*, ** Significant at 5 % and 1 % level, respectively

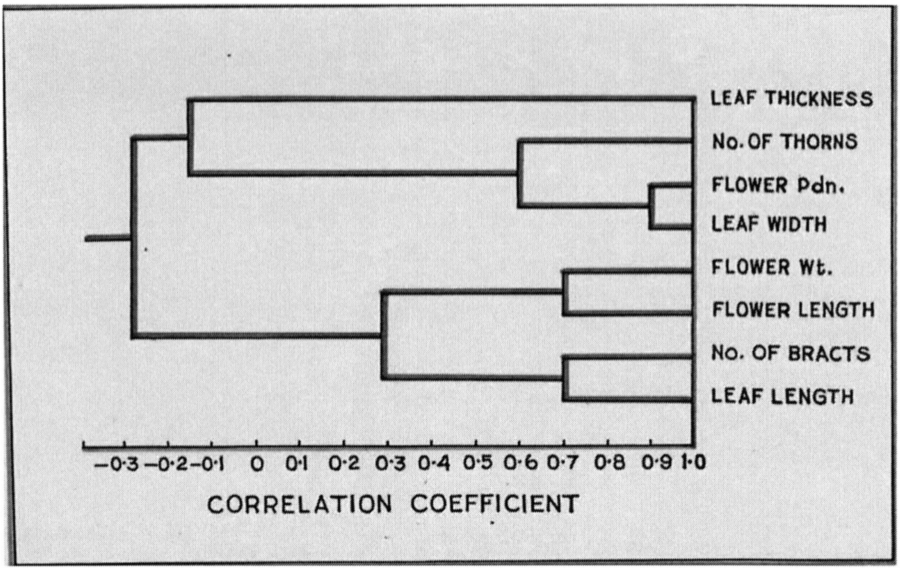

Fig. 6: Dendrogram of correlation coefficients for various traits in kewda

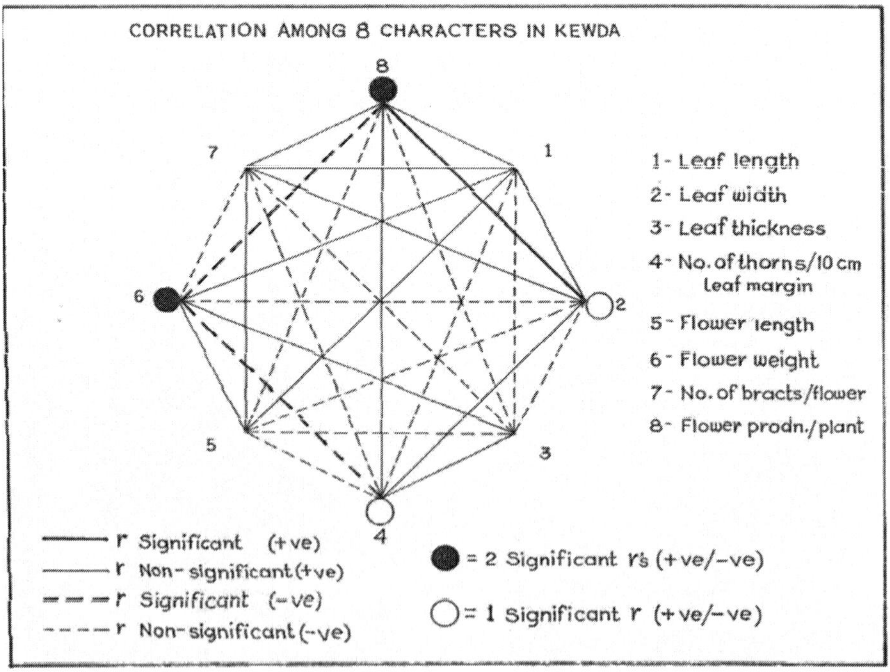

Fig. 7: Correlation diagram of various traits with flower production in kewda

Table 6: Direct (diagonal) and indirect effects of six characters on flower production at genotypic (G) phenotypic (P) levels in kewda

Character		Effects via						Total indirect effects	Correlation coefficient with flower production
		Leaf length	Leaf width	Leaf thickness	Thorns/10 cm leaf margin	Flower weight	Flower length		
Leaf length	G	0.460	0.055	0.317	-0.406	0.015	-0.056	-0.075	0.385
	P	-0.815	0.202	0.870	-0.026	-0.042	0.160	1.164	0.349
Leaf width	G	0.148	0.169	0.197	0.505	-0.058	-0.001	0.791	0.961**
	P	-0.268	0.613	0.372	0.025	0.177	-0.007	0.299	0.913**
Leaf thickness	G	-0.366	-0.084	-0.399	0.125	0.023	0.027	-0.275	-0.674
	P	0.772	-0.248	-0.919	0.014	-0.037	-0.044	0.457	-0.462
No. of thorns/10 cm	G	-0.233	0.107	-0.062	0.802	-0.066	0.060	-0.194	0.607
of leaf margin	P	0.436	0.317	-0.269	0.048	0.206	-0.189	0.501	0.549
Flower weight	G	0.088	-0.127	-0.120	0.077	-0.685	-0.053	-0.897	-0.819*
	P	-0.134	-0.425	-0.133	-0.256	-0.038	0.169	-0.561	-0.818*
Flower length	G	0.335	0.003	0.140	-0.632	-0.077	0.054	-0.100	-0.177
	P	-0.451	-0.014	0.139	-0.031	0.289	-0.150	-0.507	-0.218

Residual effect: G = -0.214, P = 0.357*, ** Significant at 5 % and 1 % level, respectively

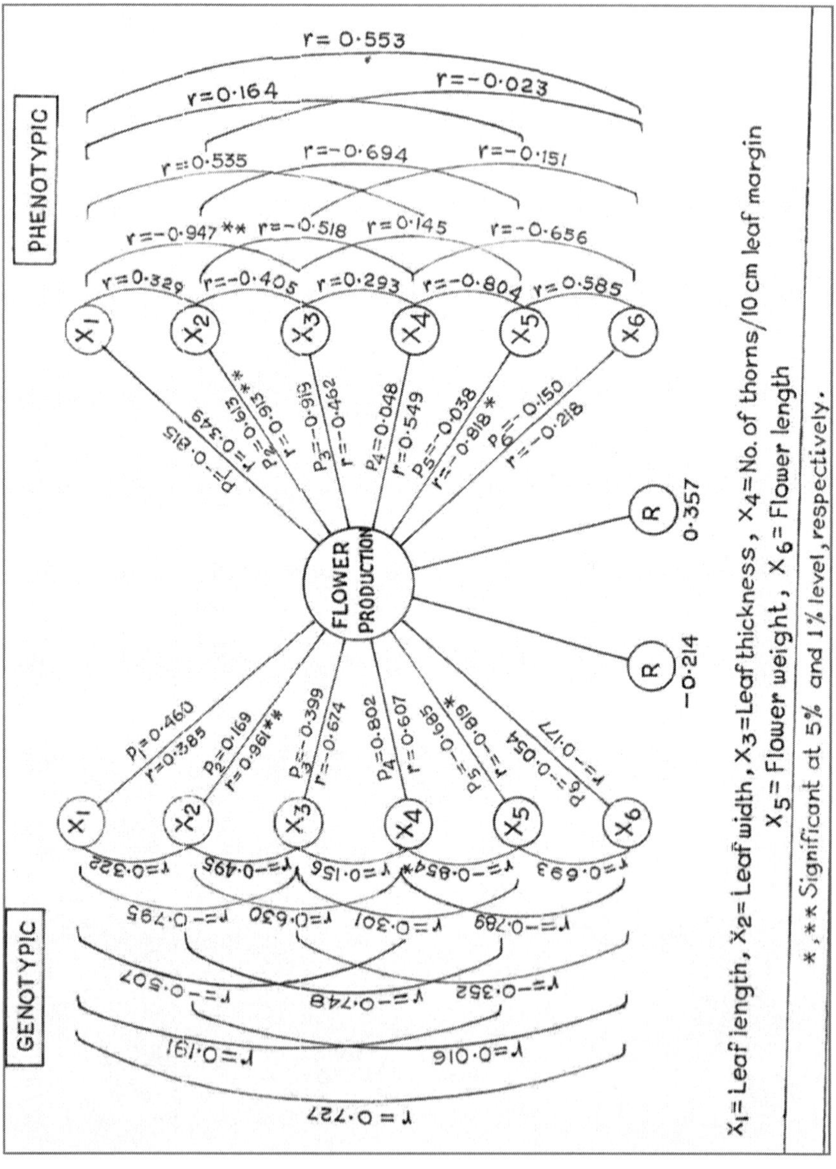

Fig. 8: Path diagram and coefficients of factors influencing flower production in kewda

Table 7: Regression coefficients (b values) of different component traits on flower production in kewda along with their standard errors, t values and linear regression equations.

Character	'b' value	Standard error	't' value	Linear regression equation (Y = flower production/ plant)
Leaf length (cm)	0.53	0.68	0.78	$Y = -16.87 + 0.53\ X_1$
Leaf width (cm)	15.09*	2.85	5.29	Y = -34.00 + 15.09 X2
Leaf thickness (mm)	-34.78	27.11	1.28	$Y = 37.46 - 34.78\ X_3$
No. of thorns/10cm leaf margin	1.09	0.77	1.41	$Y = 7.88 + 1.09X_4$
Flower weight (g)	-0.18	0.07	2.71	$Y = 37.95 - 0.18\ X_5$
Flower length (cm)	-0.23	0.50	0.45	$Y = 29.78 - 0.23\ X_6$
No. of bracts/flower	2.13	1.80	1.18	$Y = -16.91 + 2.13\ X_7$

* Significant at 5%

Selection Indices

Flower production in kewda is a complex quantitative character, influenced by variations in environmental parameters. Therefore, direct selection in populations for flower production as such may not be reliable. Consequently, selection criteria based on flower production components, would be helpful in selecting suitable plant types. Construction of selection indices based on association studies, indicated that simultaneous selection for flower weight with leaf width and (or) number of thorns on leaf margin were more efficient than straight selection based on flower production alone (Table 8) (Maharana *et al.*, 2001). Hence, the plant ideotype for higher flower production might be visualized as a plant having broader leaves with thorny margin and producing least weight flowers in kewda.

Table 8: Expected genetic advance from the use of various selection indices and their relative efficiency in kewda

Selection Index	Genetic Advance	Relative efficiency (%)
0.9 Y	18.2	100.0
14.1 X_1	18.0	98.5
1.0X_2	11.0	60.5
-0.2 X_3	15.3	83.7
2.2 X_1 + 0.8 X_2	10.6	57.9
-6.2 X_1 + 0.6 X_3	80.9	443.6
7.0 X_1 + 0.5 Y	19.7	107.9
0.1 X_2 + 0.8 X_3	73.2	401.3
1.0 X_2 + 0.9Y	25.6	140.6
0.9 X_3+1.0Y	67.5	369.8
-3.7 X_1 + 0.4 X_2 + 0.8 X_3	73.3	401.6
10.3 X_1 + 1.0 X_2 + 0.4 Y	27.3	149.4
0.1 X_2 + 0.8 X_3 + 0.8 Y	58.1	318.5
-463.6 X_1 + 20.0 X_2 + 6.6 X_3 + 44.6 Y	129.4	709.3

Y – Flower production/plant
X_1 – Leaf width
X_2 – No. of thorns / 10 cm leaf margin
X_3 – Flower weight

6

Soil and Climate

Soil

Soil has a pronounceed effect on the plant growth and productivity in kewda. Well drained sandy-loam soil with adequate supply of moisture is best suited for its vigorous growth and flower production. Thickly distributed profitableKewda plantations with normal flowering are found in the lateritic belts with deposits of silty loam soil and adequate moisture supply. In heavy clay soil with inadequate moisture supply, the plants grow vegetatively with poor flowering. At places close to the Bay of Bengal like Rangiagarh, Kujang, Paradeep of Jagatsinghpur district, where a silt-loam soil deposit is found over the clay zone, the dense plantations of kewda with free flowering ability are noticed. But surprisingly, the female plants are more at few places close to the Paradeep port, which may be due to seed propagation by drifting of ripe fruits from the uplands; while in Ganjam district, the farmers are used to plant the selected male plants in their field resulting in high flower production.

The tentative age of the natural plantations is estimated from the rates of leaf production, stem and branch growth (5-50 cm / year) and development of branching pattern (primary and secondary) in conjuction with the interactions made with the local farmers of kewda (Fig. 9). The plants of different age groups (1 to 5 years) are observed for growth characters and flower production at Tulu of Ganjam and Konark of Puri districts under farmers' field condition and it is observed that the initiation of flowers started at 5 years of age and the rate of flower production was 1.7 to 2.2 per plant (Table 9).

Besides type of soil, land situation also influenced the flower production greatly in different age groups of plants. Flower production, in tampera or saline land situation is better than dryland. Higher flower production is observed in sandyloam soil as compared to clay loam and sandy soils (Table 10). The bract length of flowers is also influenced by soil type in different locations of Ganjam district (Table 11).

Table 9: Effect of age of plantation on growth and flower production in kewda

Age of plantation (Year)	Growth characters								No. of aerial roots/ Plant		No.of flowers/ plant/year	
	Plant height (cm)		Plant girth (cm)		Leaf length (cm)		No. of branches/ plant					
	Tulu	Konark	Tulu	Konark	Tulu	Konark	Tulu	Konark	Tulu	Konark	Tulu	Konark
1	53.6	70.3	25.5	17.9	95.4	90.9	-	-	-	-	-	-
2	105.7	117.5	23.5	22.7	89.5	90.3	1.1	1.0	7.1	5.0	-	-
3	112.8	129.8	55.5	51.0	121.7	129.4	1.9	1.7	14.5	11.7	-	-
4	142.6	166.1	49.4	48.2	162.5	144.6	2.2	2.0	19.0	18.3	-	-
5	162.1	182.2	47.4	48.0	129.7	133.2	3.8	2.7	23.0	22.0	2.2	1.7

N.B: Mean data of two years

Table 10: Effect of location, soil type and land situation on flower production in different age groups of kewda during rainy season in Ganjam district

Location	Soil type	Flower production /ha (in '000)							
		Dry land situation Plant age in years				Tempera (saline land) situation Plant age in years			
		10	20	30	40	10	20	30	40
Markondi	Sandy	21.0	74.0	80.1	65.5	7.9	19.4	28.1	16.8
Tulu	Sandyloam	24.8	79.8	96.6	79.7	19.5	34.5	38.8	27.0
Panapalli	Sandyloam	23.4	79.8	95.0	72.6	16.0	29.1	38.5	23.4
Mandrajpur	Clayloam	19.9	76.0	90.0	70.2	10.5	22.3	33.1	18.4

Table 11: Effect of soil type on bract length of kewda flower at different locations of Ganjam district.

Location	Soil type	Bract length (cm) from outer to inner whorl (upto 10 bracts)									
		1	2	3	4	5	6	7	8	9	10
Panapalli	Sandy loam	61.2	49.4	40.8	33.7	29.3	26.4	24.2	21.3	20.0	18.0
Tulu	Sandy loam	50.4	41.6	35.3	30.7	26.4	23.7	21.4	19.7	17.5	15.5
Kirtipur	Sandy loam	48.3	39.8	33.1	28.2	25.2	22.0	19.2	16.5	14.0	10.9
Mandrajpur	Clay loam	23.4	19.9	12.5	11.3	10.4	9.5	8.6	7.6	7.1	5.3
Hantulu	Clay loam	16.2	14.5	13.4	12.4	10.9	10.2	9.4	8.5	7.8	6.4

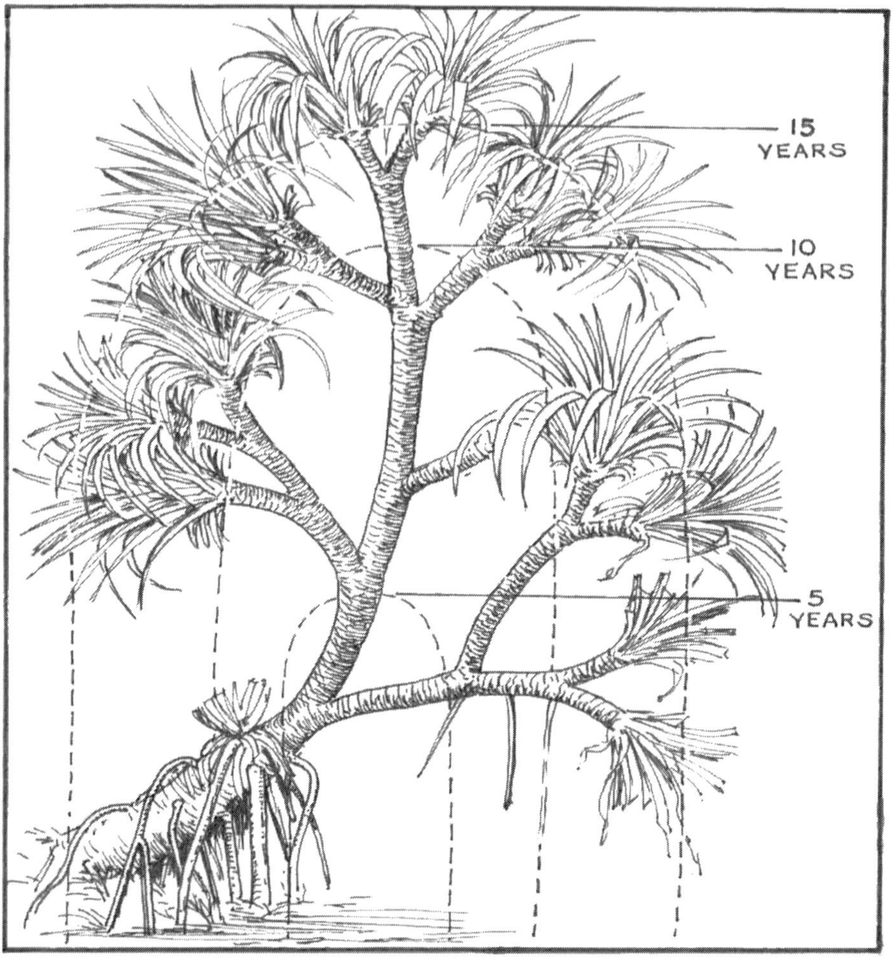

Fig. 9: Determination of age of plants in kewda

Climate

Kewda is essentially a crop of the tropics and sub-tropics requiring reasonably high rainfall, high temperature and high humidity. It can be grown at altitudes upto 300 m from mean sea level (MSL), but thrives best at places below 100 m MSL.

Plantations are well distributed and flowering is regular upto 50 m MSL and the distribution comes to a fair level at 100 m MSL and beyond 100 m, the plants are only thinly noticed on the sides of nallas, rivers and rivulets. At places upto 300m MSL, a very thin population of plants is noticed, wherever a nalla or small river runs and flowers are rarely produced. In the inland districts of Odisha like Phulbani and Keonjhar, some flowers are also produced at places with adequate moisture supply. It is interesting to note that although the plants can grow under moist conditions at altitudes beyond 300 m, but flowers are not noticed due to long cold spell and low night temperature through out the year.

The wild plantations of kewda grow along the coasts producing a belt of dense impenetrable vegetation above the high water mark. An annual rainfall of 1500 mm or more is considered desirable for kewda flower production and it is high in season of high rainfall. It is observed that the plant growth and flower production are greatly influenced by the distance from sea shore and age of plants (Maharana *et al.,* 1995)

In Ganjam district, 80% of the plantations are seen in the villages (viz. Golabandha, Keluapalli, Indrakhi, Sasanpadar, Mantidi and Viswanathpur etc.) situated within 5 km distance from seashore and rest 15 % are seen in the villages within 5-10 km distance from seashore (Kumar, 1995). Maximum flower production is noticed at places of 2 km distace from seashore in 20 years old plants and declined then with increase in distance from seashore. The increase in flower production at 30 years of age is due to the emergence of a few small flowers following the harvest of main inflorescence (Table 12). The flower production is also influenced by season and land situation. Flower production is noticed throughout the year under flowing water side (nalla site) situation, while no flower production is observed during winter and summer seasons under dryland and tampera (saline marsh land) situations. The highest flower production is observed under nalla site followed by dryland situation during the rainy season in 30 years old plantations (Table 13).

Table 12: Effect of distance from seashore on plant growth and flower production in different age groups in kewda.

Age of planting (Year)	Character values of different distances from seashore																			
	2 Km				4 Km				6 Km				8 Km				10 Km			
	Plant Height (m)	PB	SB	FP	Plant Height (m)	PB	SB	FP	Plant Height (m)	PB	SB	FP	Plant Height (m)	PB	SB	FP	Plant Height (m)	PB	SB	FP
5	1.5	2	-	5.8	1.5	4	-	5.1	1.5	5	6	4.1	1.5	5	8	3.6	1.0	6	7	2.7
10	2.0	4	12	9.7	2.5	6	12	7.8	2.0	7	12	5.9	2.0	6	14	5.4	2.0	7	12	3.1
15	2.5	5	14	14.8	2.8	7	11	13.4	2.4	9	15	11.3	2.4	8	15	9.2	2.2	8	18	6.9
20	3.0	5	15	34.7	2.8	8	18	29.8	2.5	10	18	24.9	2.4	12	18	17.4	2.3	15	20	15.7
25	3.5	7	18	31.6	3.5	8	20	27.7	3.0	14	22	22.8	2.5	18	19	15.6	2.3	20	25	13.4
30	4.5	7	20	38.2	3.5	8	25	28.3	3.1	16	25	25.6	2.8	20	25	18.5	2.5	22	30	15.4

N.B. : PB – Primary branches/plant; SB – Secondary branches/plant; FP – Flower production/plant (mean of two years)

Table 13: Effect of land situation and season on flower production in different age groups of kewda at Indrakhi region of Ganjam district.

Land situation	Season	Flower production/ Plant* in different age of plantation(years)			
		10	20	30	40
Dry land**	Rainy	8.3	25.8	35.7	26.7
Nalla site	Rainy	11.6	29.3	40.5	33.0
	Winter	4.7	4.6	6.7	5.5
	Summer	8.8	24.1	35.6	27.0
Tampera**	Rainy	8.8	15.0	18.0	12.1

* Mean of two years

** No flower production during winter and summer seasons

7

Propagation

Kewda is exclusively propagated through ground suckers and leaf bearing terminal cuttings. In kewda growing areas, predominantly the terminal shoot cuttings are used in commercial plantations. However, the seeds are also capable to germinate and give rise to plants, but it is not in practice due to hard mesocarp and seed coat (Bailey, 1950, Deshaprabhu, 1966, Mott, 1975, Maharana and Jagadev, 1993).

Seed propagation

In kewda, the flowers develop into berries, which fuse to form a multiple fruit. The receptacle and bracts are included in the fruit. The seeds are deep seated and very difficult to extract, hence seeds are sown along with the fibrous drupes. Each drupe contains 2-9 seeds. Seeds collected from ripe fruits are sown during spring. The seeds normally, take about 3 months for germination due to the hard seed coat and impervious fibrous covering. Experimentally, it was observed that one hour sulphuric acid scarification treatment of the drupes reduced the germination period by about one month increasing the germination percentage as compared to direct sowing, however, scarification for more than one hour was deterimental (Table 14) (Maharana and Jagadev, 1993). Bailey (1950) reported that seed soaking in tepid water for 24 hours would aid the germination. The seed propagated plants (Plate-27) take about 2-3 years more for flowering as compared to vegetatively propagated plants.

Vegetative propagation

Vegetative propagation is found to be more dependable for preserving and multiplying the genetically uniform selected types. Kewda is easily propagated by vegetative plant parts like ground suckers, crowns from flowering and non-flowering shoots. Success from leafless old stem cuttings (Plate 28) is very low and most of them dry in 5-6 months due to poor development of roots. Experimentally (Plate 29), it was observed that plants raised from leafless stem cuttings were died irrespective of their size (Table 15) (Jagadev *et al.,* 1993a). Ground suckers (Plate 30) were found to be the best planting material followed by the crowns (Plate 31) (Maharana and Jagadev, 1993).

As Kewda is a water loving species, its fleshy and stout stems could sustain water stress for a longer period and kewda shrubs could be shift-planted without damaging much of the top flowering buds and strong prop-roots, which mostly favour quick flowering. Shift-plantations also increase the yield and help the farmers to counter the occasional loss of flowers and measures like intercropping of kewda with aromatic plants like patchouli, vetiver and cinnamon have also been suggested (Anonymous, 2007).

Plate 27: A kewda seedling

Plate 28: Propagation by kewda stem cutting

Plate 29: Kewda research field at Konark

Table 14: Seed germination and seedling growth in kewda.

Treatment	Days to germination	Extent of germination (%)	Seedling characters at different days after sowing (DAS)														
			Plant height (cm) (DAS)			No. of leaves/ plant (DAS)			Leaf length (cm) (DAS)			Leaf width (cm) (DAS)			No. of thorns/leaf margin (DAS)		
			100	125	200	100	125	200	100	125	200	100	125	200	100	125	200
T_1	85	50	6.3	8.1	11.9	6.7	7.0	12.5	5.3	9.9	11.6	0.8	0.9	2.1	13.8	16.3	17.4
T_2	65	70	8.9	12.5	13.0	13.5	20. 5	25.5	8.3	11.6	13.2	1.0	1.2	2.2	15.0	16.0	20.7
T_3	78	40	5.2	9.0	12.5	9.0	12.0	22.0	3.7	7.1	9.8	0.7	1.1	2.1	10.5	18.0	24.0
T_4	-	-	-	-	-	-	-	-	-	-	-	-	-	-	-	-	-
T_5	-	-	-	-	-	-	-	-	-	-	-	-	-	-	—	-	-
T_6	-	-	-	-	-	-	-	-	-	-	-	-	-	-	-	-	-

N.B.: T_1 = Control; T_2 = 1 hour acid treatment; T_3 = 3 hours acid treatment; T_4 = 6 hours acid treatment; T_5 = 9 hours acid treatment; T_6 = 12 hours acid treatment *No germination in T_4, T_5 and T_6 treatments., DAS : Days after sowing

Table 15: Effect of different planting materials on survival and growth of kewda.

Character Treatment	Plant survival (%) (DAP)			Plant height (cm) (DAP)			Plant girth (cm) (DAP)		
	90	240	720	240	720	1200	240	720	1200
Crown-NF	81.3(64.5)	51.6(46.0)	46.9(43.2)	70.5	97.5	191.0	16.0	16.2	15.6
Crown-F	87.5(70.1)	57.6(46.0)	50.0(45.1)	77.1	98.1	175.8	17.1	16.7	15.7
Stem cutting (0.5m)	28.1(32.0)	-	-	-	-	-	-	-	-
Stem cutting (1.0m)	29.7(32.9)	6.3(14.5)	-	74.0	-	-	21.0	-	-
Ground sucker	84.4(67.3)	75.0(60.6)	73.4(59.7)	71.1	100.1	165.5	14.2	13.7	13.6
SEm (±)	3.9	5.7	5.5	6.4	6.3	10.2	1.3	1.3	1.1
CD (0.05)	8.4	13.0	13.5	N S	N S	N S	2.9	N S	N S

CharacterTreatments	No. of new leaves/ plant (DAP)			No. of aerial roots/ plant (DAP)			No. of side suckers/ plant (DAP)		
	240	720	1200	240	720	1200	240	720	1200
Crown-NF	6.9	11.1	7.4	0.6	4.9	7.4	2.5	3.6	8.0
Crown-F	6.8	9.9	7.7	1.0	5.7	9.6	1.9	2.5	7.6
Stem cutting (0.5m)	-	-	-	-	-	-	-	-	-
Stem cutting (1.0m)	6.0	-	-	-	-	-	-	-	-
Ground sucker	6.8	10.1	7.9	1.0	5.7	7.4	1.3	2.2	7.3
SEm (±)	0.3	1.2	0.4	0.4	0.5	1.5	0.8	0.6	0.7
CD (0.05)	NS	NS	NS	NS	NS	NS	NS	NS	NS

Figures in parentheses are transformed angular values
DAP = Days after planting, NF=Non-flowering part, F=Flowering part

Plate 30: Propagation by kewda ground sucker

Plate 31: Propagation by kewda crown from flowering part

8

Cultivation

In practice, flowers are collected from wild plantations. Recently kewda has emerged as one of the important commercial commodities. A systematic cultivation practice is essential to develop a successful plantation and to increase the productivity.

Planting

Kewda suckers or crowns are generally planted during June-July. i.e. in the beginning of the monsoon, when the soil is not too wet. Planting in wet soil delayd the root formation and reduces the success level. Just after the first shower of rain or during the summer shower close to monsoon, kewda is planted.

In a full grown plant, stilt roots originate at a few feet above the ground to render a support for the huge canopy of the plant. Each part of the stem is capable enough for development of root initials, but it is retarded once the stem became old and leafless. The root initials develop at the terminal portion, where more leaves cluster, but they do not grow. Hence, the stem cuttings collected from the flowering or non-flowering shoots bear the rudimentary roots and when planted, they establish quickly.

In practice, the cuttings are collected and scattered on the ground for two to three days. Before planting, a few leaves closed to the base of the cuttings are removed to expose the root development sites. The cuttings are planted in trenches of 45 cm depth horizontally or at an angle of 30^{0}. The new leaves appear on the growing points one month after planting.

Kewda plants are usually planted on the field bunds for fencing and subsequently used for collection of flowers. A 30 years old plant can produce a crown of 6-7 m, for which it is difficult to develop a commercial plantation in a close spacing. It has been observed in the natural plantations that the plants may be spaced at 8-10 m distance to achieve a full length of plantation life. It is, therefore, suggested by the authors to follow a paired row method of planting maintaining a spacing of 4 m between plants and rows of a paired row and 6 m between two paired rows; which provides all facilities for plant growth upto 10-12 years and then pruning of a few branches are required. Hence, a plant population of 500 plants in the field and 150-200 plants on field bunds are maintained in a land of one hectare.

Manuring and Fertilization

Practice of manuring and fertilization is hardly followed in kewda plantation. Kewda can be grown in a variety of soils, but sandy or sandy-loam in texture is good for its cultivation. However, each plant may be dressed with 1-2 baskets of FYM or organic waste compost before planting. Experimentally, it has been found that kewda responds to application of inorganic fertilizers. Annual application of 60 g N, 30 g $P_2O_5$and 30g K_2O per plant in June (i.e. just before monsoon) is beneficial. The recommended fertilizer dose per hectare is 30:15:15 kg NPK.

Weeding

Weeds are not a problem for its plantation because of salinity and littoral type of soil in kewda growing areas. Weeds do not grow, but plantations closed to the canal embankment and field bunds compete with some weeds during the initial stages of growth. Once the plant growth peaks up, weed growth is automatically suppressed due to shade and development of extensive root system. Hence, no practice of weeding is followed.

Intercropping

In kewda growing areas, some wild melons grow and are highly compatible with kewda. Since kewda takes more than 5 years to flower, some legumes and melons can be grown as intercrops because the soil condition is unsuitable for the growth of any other crop. Intercropping keeps the soil loose and supplies nitrogen to the crop.

Water Requirement

Kewda requires a plenty of water for its growth and development, but tolerant to drought, where the water table is very high. The crown portion dries during summer and favours fungal growth during rainy season. It is, therefore, desirable to provide some water during summer to sustain the vegetative growth. Kewda grows luxuriantly in soils having high water table. Soils deficient with moisture do not allow to produce stilt roots and as a result the crown growth is suppressed and favours production of suckers. In the drier inland districts of Orissa, the plants don't produce stems and remain bushy throughout. Flowering is only noticed in the plants with much branches in the areas having high water table and continuous water supply favours flower production. The places receiving rain for a longer period give a continuous harvest of flowers. The flowering season only coincides with rainy days and moderately to winter.

Pruning

Kewda produces a plenty of suckers. Hence, if unpruned, the plant forms a bush with large number of suckers or low growing branches. Suckers are usually produced after one year of planting. To facilitate the linear growth, desuckering should be done three years after planting and it should be practised at least once a year subsequently to give a particular direction of growth to the plants. Pruning of the old and undesirable branches of the old plants is advisable to maintain vigour and good quality flowers production.

Flowering, Pollination and Fruit Set

Like all *Pandanus* species, kewda is also dioecious and bears male and female flowers on separate trees (Hyndman, 1984). Female flowers do not have any characteristic smell, but bear aggregate fruits comprising of numerous ovaries that develop into drupes. Spadix of male flowers (25-30 cm long) with several cylindrical spikes of 5-10 cm long are enclosed in long creamy white fragrant spathes (Misra and Rao, 1997).

In kewda, vegetatively propagated plants come to flower after five years of planting, while seed plants take about seven to eight years. A fully grown plant (15-20 years old) yield about 15-25 flowers in rainy season, while older plants (25-30 years old) bear upto 35 flowers. Most plants show an increase in number of flowers that come each year from an age of 15 years to 30 years and then gradually decline due to loss of branches lasting the flowering phase upto 40 years or more. Flowering period extends from July to October and a very few flowers come during the winter months. Male plants form about 30 branches and flower annually, but female plants form about 16 branches and flower biennially. For plants of same age, the males produce about three times as many flowers as the females (Ash, 1987).

Application of growth substance has proved effective in encouraging early flowering and production. Spraying of planofix (α-naphthalene acetic acid formulation) at floral bud initiation stage at a concentration of 50ppm during normal season and 25 ppm during off-season of flowering induces early and more number of flowers in kewda (Jagadev et al., 1999/2000)(Table 16).

Table 16: Effect of planofix on flower production and its attributing traits in kewda

Treatment	Days taken to flower production (DAT)	No. of flowers/ plant	Weight of flower (g)	Length of flower (cm)	No. of bracts / flower
Normal Season					
Control	NA	10.0	100.0	27.2	16.3
50 ppm	NA	10.6	115.1	29.6	13.5
Off-Season					
Control	43	3	103.6	27.7	16.6
25 ppm	40	8	78.7	30.9	16.6
50 ppm	36	5	85.5	27.5	15.0
100 ppm	50	6	75.0	26.5	14.3

N.B. : DAT = Days after treatment NA = Not available

The pollination in kewda is anemophily as the pollen grains are light and produced in abundance on the androecious fesicles. The pollination, possibly is also assisted by insects (Purseglove, 1975).

The fruit set occurs during rainy season, when the temperature is low and the relative humidity is higher. Fruit takes about one year to mature . If, however, female flower fails to set fruit, it is possible the plant may flower again in the next year. A failure in pollination or fruit set in one season may, therefore, induce synchronous flowering in most females by next year and establish a synchronous biennial pattern. (Ash, 1987). Each fruit forms about 75-80 phalanges (drupes) and each phalange contains up to nine embryos. As kewda is dioecious, the dioecism may have been maladaptive on long distance dispersal, but it has been largely overcome through the evolution of facultative apomixis (Cox, 1990). In the absence of pollination, it develops embryos through apomix is upto 18% and can act against the development of a synchronous biennial pattern of female flowing (Ash, 1987).

9

Harvesting and Yield of Flower

Kewda plant occurs wild in the entire coastal belt of Odisha , but abundant in Chhatrapur and Keluapalli areas of Ganjam district. There were about 3-4 lakhs trees of kewda in Ganjam district (Deshaprabhu, 1966). Main season of flowering starts from July-August and lasts for 2-3 months and a few flowers also come during November-December and March-May (Haines, 1961). Averagely, a plant produces 15-20 flowers per season depending on branching of the tree (Plate 32). A fully matured tree bears 30-40 spadices in a year (Deshaprabhu, 1966). Flowering is also dependant on rainfall, being better in years of high rainfall. Size of flower is bigger during rainy season. The spadices take a fortnight to mature, depending upon the weather condition. Ripe spadices, creamy in colour, give higher yield of perfume and of better quality than immature ones (Plate 33). The kewda flower of Ganjam district is also registered under the geographical indications (GI) of goods (Registration and Protection) act (Panigrahy, 2019).

The flowers are mostly collected from the natural plantations within a radius of 15 km around Chhatrapur, Tulu and Keluapalli areas. Generally, 60 % of flowers collected between July to September, 30 % between May to June and about 10 % between October to November from the plantations for the distilleries (Padhi, 2017). Plucking of flowers is hardly made by hand, but always with a hooked stick (Plate 34). Collection by cutting down flowering stalk is not recommended as it may damage the floral primordia and arrest further flowering. The labourers are engaged by the local people to collect the flowers and it is observed that in Tulu region 60% of the flowers are collected by male, 30 % by female and rest 10% by child labourers. But the growers used to pluck the flowers from the field by engaging their family members of which 70 % by male, 25 % by female and rest 5% by child members. Collection of flowers is completed by early morning, so that those can be carried in baskets to the distilleries (Plate 35) by 9.00 a.m. for distillation on the same day as longer storage results in deterioration of oil quality. The collection of flowers from natural plantations can be considered as a family business, since most of the family members from children to the adults are engaged in collection during the early morning hours and individuals with headload of flowers, even in small quantity start moving to the distilleries.

Sometimes middlemen also work for their supply to the distilleries. It is observed that 83 % of the flowers picked are carried by the people on headloads, 15 % by bicycle and 2% by other vehicles like jeep. At the factory site, the collected flowers are counted by a literate person, employed by the owner and recorded in the daily receipt register from 7.30 a.m. to 10.00 a.m. The collected flowers need to be sold on the same day, or else the flowers will be of no use for distillation due to the loss of moisture in transpirational process (Anonymous, 2011). The flowers delivered to the factory is spread in a cool store house. The height of the flowers heap sometimes reach almost to 1.0-1.5 m.

Plate 32: Kewda plants in full flowering stage

Plate 33: Kewda male flower at different stages of development

Plate 34: Kewda flower collection with the help of a stick (www.whitelotusaromatics.com)

Plate 35: Kewda flowers being carried over head for distillation (www.whitelotusaromatics.com)

Generally, for procurement of flowers, the factory owners advance to the growers during the off-season for supply of flowers in flowering season at a constant price and the price of the flowers irrespective of their size is fixed every year by a committee formed by the local growers which also floctuates depending on demand and supply. Even today, the kewda growers sometimes reach at the door of the attar manufacturers for selling of their collected flowers as yet a kewda corporation has not been formed and the price of flower is not fixed. The poor farmers around Chhatrapur area do not get an alternative other than the kewda plantations as the paddy crop cultivation is not profitable due to the soil salinity. The annual flower production is about 36 millions of worth about Rs.18 crores and the major share of this income (about 95 %) goes to 154 villages of Chhatrapur and Rangeilunda blocks (Anonymous, 2007).

Now-a-days, the price of the flower is increased to many fold i.e. Rs. 0.60 in 1986, Rs. 1.35 in 1992, Rs. 2.10 in 1993, Rs. 5.00 in 1994 and Rs. 7.00 in 1995 and Rs. 9.00 in 1996, Rs.10.00 in 2000 and Rs.15.00 in 2006 (Anonymous, 2008), but now the price of flower decreased to Rs.8.00 (Padhi, 2017). The average weight of a flower varies between 90-100 g and it is estimated that about 30 million flowers (3000 tonnes) are distilled annually in the Ganjam district (Dutta *et al.,* 1987; Misra and Rao, 1997), although the potential yield may be much more as a large number of flowers remain uncollected from many areas which are not easily accesible. As many of the flowers bloom in the densed impenetrable plantations, only 15% of the total flowers that bloom are utilized for extraction (Anonymous, 1995) Sometimes plantations are cleaned to provide passage for higher collection of flowers. As the price of a flower ranged from Rs.4/- to Rs.10/-, a farmer having 10 kewda shrubs can earn an additional amount of Rs.5,000/- to Rs.8,000/- per month by selling 20-25 flowers per day (Subudhi, 2012).

10

Chemical Composition

The volatile oil in the flower determines the odour and flavour of kewda and it has been of particular fascination to the chemists. There are only a few reports on the chemical composition of kewda essential oil (*rooh* or *otto*) Dhingra *et al.* (1954) distilled the kewda flowers collected from five places of Ganjam (Odisha) and Ghazipur (Uttar Pradesh) districts. He found the oil yield to vary from 0.015 to 0.031% and the compounds identified were methyl ether of β-phenyl ethyl alcohol (65.97%), d-linalool (18.70%), dipentene (8.31%), phenyl ethyl acetate (3.48%) and citral (1.82%). The oil was of light to golden yellow in colour having the following physico-chemical properties *viz.*, specific gravity (d20°) 0.932-0.958; refractive index (n20°) 1.485-1.491; optical rotation ([α]D) + 1.35° to + 2.08°; acid value 1.4 – 6.7; ester value 15.9 – 40.9 and soluble in equal volume of 80 % alcohol (Deshaprabhu, 1966). Sadgopal (1959) also reported the major constituent of oil as methyl ether of β-phenyl ethyl alcohol (65 – 80%) and minor constituents as dipentene, α-linalool, phenyl ethyl alcohol, citral and some terpene esters.

According to Chandra (1985), the oil contained dipentene (6.0%), d -linalool (19.0%), phenyl ethyl acetate (4.6%), citral (1.8%), caproic acid (0.5%) and stearoptene (0.2%) besides the major constituent methyl ether of β-phenyl ethyl alcohol (67.0%).

On the basis of gas chromatography, the oil was found to contain methyl-β-ethyl ether (74.67%), β-phenyl ethyl alcohol (16.23%) and phenyl ethyl acetate (2.76%) as the major constituents along with numerous other trace constituents like nerol, neryl acetate, citronellyl acetate, farnesol and nerolidol (Nigam and Ahmad, 1992). Naqvi *et al.,* (1995) also examined 10 samples by GLC and found the major constituents as phenyl ethyl methyl ether (75-84%) and terpinen-4-ol (8-15%) with optical rotation (+ 1.10 to + 2.8°) and refractive index (1.48-1.49). They concluded that a good quality kewda oil should have optical rotation between +2.00 to +2.80 and terpinen-4-ol not less than 10.0%.

By capillary gas chromatography of kewda oil, Maheshwari (1995) showed the presence of five major components *viz.* methyl ether of β-phenyl ethyl alcohol

(71.08%), terpinen-4-ol (17.72%), α-terpeniol (1.79%), γ – terpenene (1.87%) and cineole 1:8 (0.87%). Two major components viz. β-phenyl ethyl alcohol (16.2%) and phenyl ethyl acetate (2.76%) reported earlier by Nigam and Ahmad (1992) had been found to be present only in traces. Out of 55 peaks, 43 were identified and 28 being reported for the first time.

By GC-MS analysis of the two typical genuine samples of kewda oil collected from Ganjam district, Bisht *et al.* (1997) reported that the oil contained more than 50 constituents, of which 44 were identified and 20 being reported for the first time. The major constituents of the oil were methyl ether of β-phenyl ethyl alcohol (67.13%) and terpinen-4-ol (19.48%) followed by α- terpineol (1.97%), p-cymene (1.25%), 1:8 – cineole ((0.76%), γ- terpinene (0.52%) cis- and trans-p-menth-2-en-1-ol (0.48% and 0.35%), α-pinene (0.35%), limonene (0.30%) and other minor components. Some of the compounds reported earlier in large quantities, were then shown to occur in traces or absent. The major fragrance component of the flowers is methyl ether of γ-phenyl ethyl alcohol, popularly known as 'kewda ether'. Misra and Rao (1997) also reported four new comopounds, *viz.* β-terpinene (0.93%), trans-menth-2-en-7-ol (1.28%), trans-α-ocimene (1.02%) and 2-methyl-6-methylene-7-octen-2-ol (2.10%) by GC-MS study of kewda oil.

Misra *et al.* (2000) reported that the major components of essential oil, water distilled from staminate inflorescences of kewda were 2-phenethyl methyl ether (65.6-75.4%), terpin-en-4-ol (11.7-19.5%), p-cymene (1.0-3.1%) and α-terpineol (1.2 – 2.9%) and of ketaki were 2-phenethyl methyl ether (66.2%), 4-hydroxy phenol (6.1%), 2-phenethyl alcohol (4.5%), 3-acetyl anisole (2.26%) and p-benzoquinone (1.3%).

From the above literature on chemical composition, it was inferred that the major fragrance component of the flower was methyl ether of β-phenyl ethyl alcohol. However, for a long period of time phenyl ethyl alcohol was considered as a second major component of kewda oil, but it was not true. In fact, it was terpinen-4-ol as evident from GC-MS. The erroneous inference drawn in the characterization of constituents by earlier researchers, was due to the non-availability of capillary GC. The natural components present in kewda oil play important role to impart a characteristic powerful kewda odour or flavour. If any synthetic chemical was added, it changed the odour or flavour of the oil greatly.

The fruit of kewda is also edible and when ripe, becomes sweet smelling and orange-red in colour. Platt (1962) reported that the edible portion of kewda fruit contained moisture (80.0%), carbohydrate (19.0%), protein (0.4%), fat (0.3%) and fibre (0.3%). The authors reported earlier that the ripe kewda fruit collected from Ganjam district contained moisture (80.00%), pulp (18.00%), crude protein (0.87%), total soluble solid (15.00%) having reducing sugar (0.14%) and acidity (0.38%) (Maharana and Jagadev, 1993). Hence, the fleshy base of the drupes can be processed into different preserved products like squash, jelly and slices in syrup (jam) etc.after neutralizing certain irritant raphides present in it by PVP (polyvinyl pyrolidone) like the clarification of cashew apple juice.

11

Insect Pests and Diseases

Kewda plants are hardly attacked by the insect pests and diseases. However, a few fungal and bacterial diseases are found to reduce the yield of flowers significantly. Some important insect pests and diseases of this plant with their nature of damage and suggested control measures are discussed here as reported by Jagadev *et al*. (2001).

Pests

1. Bag worm

During the dry season, leaves are found to be attacked by bag worm (*Manatha albipes*; Fam. Psychidae; order- Lepidoptera) (Plate 36). The larva, a foliage feeder, feeds on the chlorophyll and veins of leaves leaving the mid-rib resulting in perforation of leaves. In case of heavy infestation, the leaves become dry. Extensive damage arrests the vegetative growth resulting in significant reduction of flowering. Thus, the insect is considered to be of economic importance, but till date neither the extent of economic loss caused to the crop nor its control measures have been worked out.

Suggested Control Measures

Spraying of any contact insecticide e.g. Endosulfan or monocrotophos @ 0.05% a.i. is suggested at an inteval of 15 days if the population of insects is noticed on the plants.

2. Beetles and Thrips

Some saprophagous beetles (Fam.-Chrysomelidae; Order. Coleoptera) and thrips (Fam. – Thripidae; Ord. – Thysanoptera) are observed on the inner side of bracts and on the androecious spikes of male spadices. Although, the beetles are found feeding on the pollens of the flower, but at times these cut the flower bracts and their excreta also deteriorates the quality of flower. Further research to know their role as a pollinator or causing any harmful effect to the crop will be fruitful.

No control measure is suggested for these insects. A predatory lady bird beetle (*Micraspis discolor*; Fam. – Coccinellidae; Order - Coleoptera) is also found on the leaf which feeds upon small insects like thrips and aphids.

Diseases

1. Foot rot of Central shoot

During the summer months, the foot rot of central shoot of kewda is noticed in some plantations and the extent of damage is upto 15%. The disease is characterised by the presence of water soaked, discoloured leaves in the whorl and it is more pronounced at the basal portion. Diseased leaves can be pulled out easily from the bunch and emergence of new leaves are hampered or stopped. In advanced stage, the diseased leaves with necrotic lesions show the presence of slimy bacterial oozing and leads to premature drying of plants. The causal agent is *Pseudomonas solanacearum* (Plate 37).

Suggested Control Measures

Preventive measures are to be taken by mixing the soil with stable bleaching powder @ 12 kg/ha before 15 days of planting or the planting materials should be treated with plantomycin @ 0.2% solution for 15 minutes.

2. Leaf blight

During rainy season, the disease is mostly observed on the leaves and the extent of damage is 5-7 %. The disease is characterised by extensive yellowing of leaves. The large spots turn deep brown and often cover the entire leaf leading to premature defoliation and significant reduction in flowering. The symptoms are mostly exhibited under high humidity and high temperature (Plate 38). The causal agent is reported to be a fungus, *Botryodiplodia theobromae* (Suryanarayanan and Narasimhan, 1981). It is also reported to be caused by the fungus, *Alternaria tenuis* Nees (Purseglove, 1975).

Plate 36: Kewda plants attacked by bag worm (*Manatha albipes*)

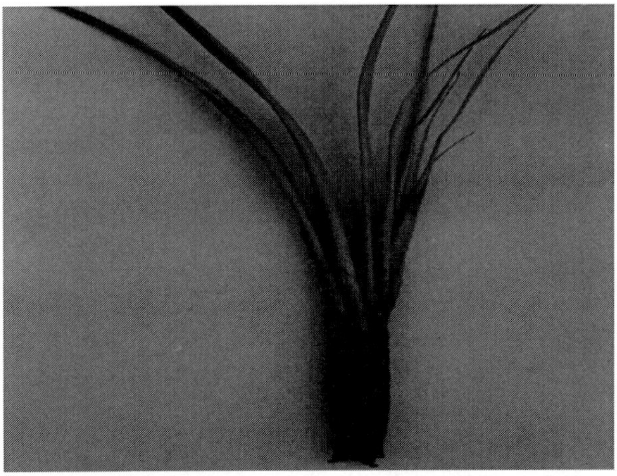

Plate 37: Foot rot disease of central shoot in kewda

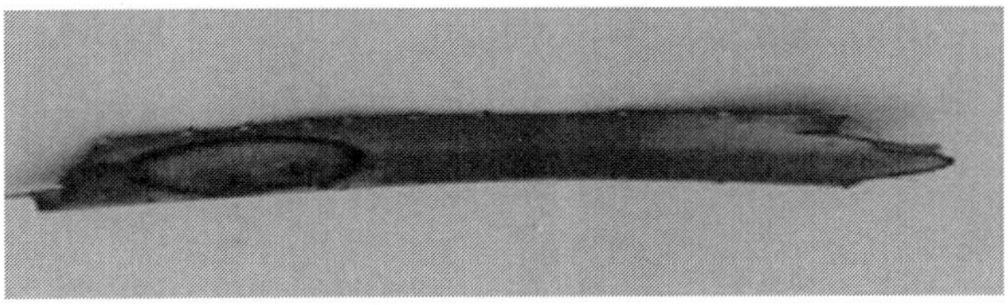

Plate 38: Leaf blight disease of kewda

Suggested control measures

Two sprayings of Dithane M-45 @ 0.4% or Blitox 50 @ 0.2% or Bavistin @ 0.15% at an interval of 10-15 days is suggested.

3. Fruit rot

The fruit is rarely eaten by the local folk. It can also be used for food processing at ripening stage. Sometimes, just before or at ripening stage, the central core rotting of fruit is noticed during the rainy season, resulting in premature dropping of fruits.

Suggested Control Measures

The planting materials for propagation should be collected from disease free plants to reduce the incidence of disease.

Besides the above, the occurrence of location specific diseases are also reported in screwpine. From Papua New Guinea, Tomlinson (1988) reported that the leaf necrosis and fruit rot diseases in *Pandanus conoideus* are caused by *Erwinia carotovora,* subsp. *carotovora.* From Satara (Maharashtra), Chavan and Hosagoudar (1984) identified a new fungus, *Vestergrenia pandani* in *Pandanus facicularis (kewda).*

12

Perfume Production

French author Jean Giono once remarked, "The God creates the Odours, but the men manufacture the perfumes". In today's world, fragrances play bigger role than just being associated with a tiny bottle of liquid. It now serves the larger purpose of linking itself with anything you can imagine, from soaps to your favourite sweater knitted by your grandmother (Biswas, 2012). In the modern day, these products find their applications in items for personal sophistication to culinary delights and beverages, to a host of industrial products (Dev, 1993). India, a country known for its vast range of fragrant flowers and herbs, has used extracts from such raw material to produce incense sticks, medicines, body massage oils and sweet savouries.

The word 'perfume' is derived from the Latin term 'perfumen', meaning 'through smoke'. *Perfume* is a fragrant liquid mainly distilled from natural oils, aromatic chemicals and resinoids; *cologne* is a lightly scented liquid used to cool the skin and *deodorant* contains chemicals to counter offensive body odour. Perfume is basically used by women, cologne by men and deodorant by both (Saha and Sareen, 1997). Throughout antiquity there was a continuous demand for incense, to mark the rites of scenting passage from birth to the grave, for the daily rituals of incensing rooms, linen and clothing. The South Asian synonym for perfume, 'attar', on the other hand, is derived from the Arabic word *'aettr'* meaning 'fat' or 'extract'. Almost all attars are basically aromatic oils extracted from flowers, plants and herbs through steam distillation. Some fragile flowers and oils that can not stand the heat of distillation are, however, crushed and the oil expelled by mechanical force (Mahmud, 1992).

Origin and History of Perfumes

The story of perfume began around the year 4000 B.C., as early as the first flowering of the great civilizations of the world. But where did it all start ? China, Sumeria or at Mahenjodaro in India ? Anthropologists and historians can exchange radio carbon dates until they are blue in the face about the origins of civilization, but the fact remains that perfumes have, through the ages, represented the very texture of life with al its religious, medicinal and scientific ramifications (Mahmud, 1992).

The history of perfume stretches back to the ancient Egyptians, who used fragrant wood and essence at their temples as offerings to God. In the renaissance, perfume became a symbol of luxury and prestige for wealthy Europeans and the new merchant class. The end of French revolution saw the birth of the first major perfume houses in Paris to serve a fresh clientele-the new middle class. In the 19th century, perfume grew rapidly in popularity due to new advances in chemical synthesis (Anonymous, 1995).

Some olfactory facts mentioned that Cleopatra had the sails of her ship dipped in perfume before venturing out and when she welcomed Mark Antony to her boudoir, her bed and floor were shown with roses. Perfumes were Cleopatra's strongest tool to seduce Mark Antony. The ancient Egyptians used perfumes as offerings to their deities; they occasionally used it for aesthetic purposes (Walia, 1997). Emperor Nero had silver pipes installed to spray his dinner guests with rose water and once paid the equivalent of US $ 1,00,000 for a water fall of petals which smothered one guest and killed him. Victonians were afraid of the sexual powers of perfume and did not allow it to touch their skin, instead they scented handkerchiefs and slippers (Anonymous, 1995).

The land of perfumes, France, has very interesting anecdotes about the origin of perfumes. It appears that during the reign of king Louis XIV, XV and XVI in Vsersailles, the use of ordinary water for bathing and drinking was considered to be taboo. To eliminate the unpleasant odour of perspiration, an extract from the bark of a fragrant tree was applied liberally all over the body. Madame Pompadour spent the equivalent of more than a US $ 2,50,000 on perfumes and balsams to impress her royal lover, Louis XV with true orgies of scent. It was during this era that the perfume industry was born (Dey, 1995). Perfumery has its origins in France, where fragrant houses were created to market the fragrances. These fragrance houses were proprietorial in nature and the owner was the perfumer (Biswas, 2012).

Selection of a Perfume

In the western culture, gifting perfumes to ladies is considered very romantic. This is specially true for occasions such as birth days and anniversaries. Factors assisting selection of a perfume could vary from climatic conditions to intensity of body odour to the mood of a person. For people on the move, a moderately strong and long lasting fragrance is suitable. During formal occasions, a perfume which is mild yet pleasant to the senses is preferred. At times, the personality of the individual can also help in selection – charismatic or flamboyant persons tend to use strong fragrances as against the sophisticated and low profile person, who may prefer a mild fragrance.

Ancient Indian Perfume Industry

India's perfumery tradition can be dated back to over 5000 years to the Indus Valley civilization. From the excavations of Harappa and Mahenjodaro, a water distillation still and receiver have been recovered which are similar to *deg* and *bhabka* used in distillation units. In the days of Roman trade with India, Indian sandal wood, Kashmiri saffron, Assami black agar, Himalayan kasturi (musk), several essential oils and different floral attars were regarded as requisites in the toilet of a Roman gentleman.

In the most ancient Indian civilization of Mahenjodaro, aromatic resins and sandal wood were used primarily for religious rites. Massaging the body of a bride-to-be with *ubton*, a paste of sandalwood and turmeric, is an ancient custom that continues till today (Mahmud,1992).

During the reign of Harsha Vardhan (606-647 A.D) the perfume industry took a firm footing and the trade flourished thereafter. Taking advantage of the flourishing trade in perfumes way back in the Harsha Vardhan period, a tax was levied on khus and so continues the tradition to date of taxing heavily the perfumery products by the Government. The essential oils and fragrances from India entered Europe through Egypt in ancient and medieval times under the name of Arabian perfumes and Babylon. Greek, Rome, Persia and Iran were the exporters of these valuable foreign exchange earners.

The Moghul emperors were great connoisseurs of exotic perfumes. Abul Fazl in *Ain-e-Akbari* mentioned that his majesty Emperor Akbar (1547-1605 A.D.) was exceedingly fond of perfumes and the court chamber was continuously scented with flowers and fumigated with preparations of Ambergris and Aloes in gold and silver censers.

Emperor Jahangir (1605-1625 A.D) in *Tojak Jahangiri* referred to distillation of rose water. His empress Noor Jehan had the habit of taking baths in tanks filled with warm water and rose petals. On a cold morning in 1612 A.D. she noticed an oily substance floating on water, which was nothing but the present day *rooh* rose (absolute) and she was reputed to have developed the process of preparation of *'attar'* from roses as initially the perfumery products were used as such without isolation of the odorous principles. Since that time the 'attar' industry grew in and around Kannauj in Uttar Pradesh and slowly spread all over India (Sachdeva, 1993).

Our country was one of the origins of natural perfumes called 'attars'. This world's first natural perfume, was used and also exported from India. This industry in fact, was flourished and reached its peak in Moghul era. In spite of, such historic and glorious past, we are now far behind in the competitive world market of fine fragrances.

Making natural oil is not an easy task as it is a very costly affair. The manufacturing cost of the product is around 30-35% and there is still have a margin of 60-65%, out of which about 18 % goes to GST and 40% is eaten up by the distributors and retailers roughly. Manufacturing the natural oil blends is an expensive affair. The price of some of the natural oils at the raw stage is mind boggling. Six million flowers produce one litre of jasmine oil. The quantity used to manufacture the raw material is proportionally too high and therefore the price too is equally high. As regards to world production, it has been observed that developing countries contribute about 55 %, industrialized countries 35 % and European countries produce about 10 % of the essential oils and the countries for fragrance indicate that European countries use about 37 %, American countries 36 %, Asian and Oceania countries 23 % and rest countires of the world use about 4 % of the essential oils. If we see the essential oil use pattern, about 30 % used in household products, 23 % in soaps and 10 % used in oral hygiene products. In kewda perfumes use pattern, about 75 % is used in pan masala and chewing tobacco industry and rest 25 % used for other purposes (Fig. 10). Total world production of essential oils was around 1.0 to 1.1 lakh tonnes and India stands 3rd with a share of 16-17 %, but in value terms, stands at 2nd with a share of 21-22 %. Brazil is the largest producer of essential oils due to its production of citrus oil at 40,000 tonnes with 90 % share in value terms, but USA is the largest consumer of essential oils (Farooqi and Sreeramu, 2001).

A litre of rose oil costs Rs. 1-3 lakhs and *Eucalyptus* oil is Rs. 400 a litre. On an average, the per cent of essence in the plant varies between 0.1-0.3%; i.e. 80 kg of rose flowers are required to obtain one ounce of rose oil (Sahay, 1998) or 60,000 rose flowers to produce one ounce of pure rose oil (Manral, 1996)

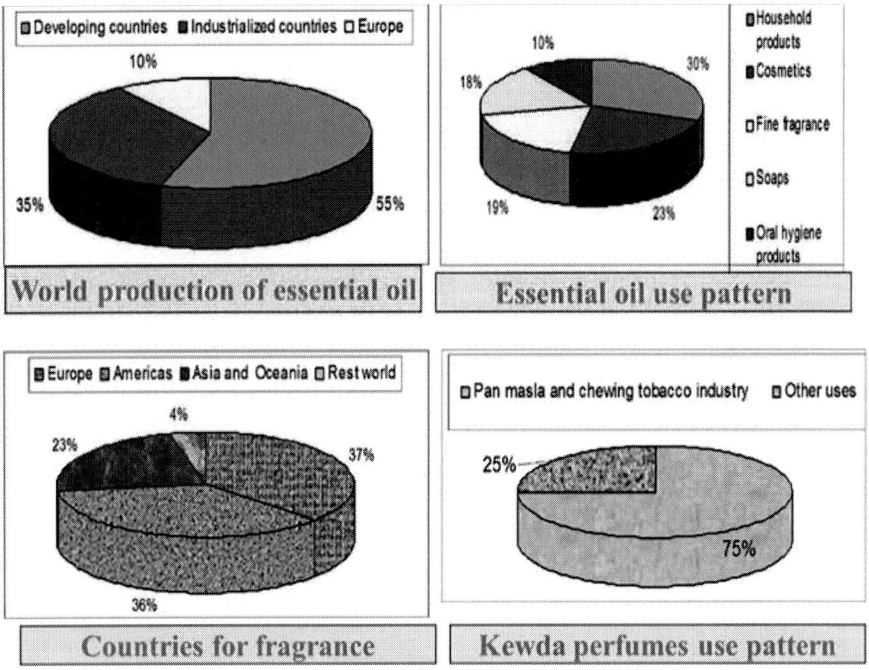

Fig. 10: World production of essential oils and kewda perfumes use pattern

Modern Indian Perfume Industry

The perfume industry in India has acquired considerable importance world over, as many essential oils like sandal wood, palmarosa, mint, citronella, ginger grass, lemon grass, *Eucalyptus,* cedar wood, turpentine, camphor besides the traditional attars like rose, kewda, khus, mogra, genda, mehindi and hina are produced here. In India, the modern flower industry had a modest beginning at the turn of this century at which point of time flavours were imported. The manufacture of flavours began on a small-scale only during the 1940s and today. India is known world over for its attars. Our perfumers have blended fragrances to match the most fancied foreign brands like Opium, Poison and Charlie etc. The Indian perfumers, if not as good as the French, but are one among the best in the world.

Perfumes in India

The immense potential of perfumes in the Indian market has coerced many international companies to enter India. The perfume industry in India has begun to look up with the advent of world renowned brand names vying for the maximum possible market share. Famous international brands such as Yardely, Nina Ricci, Fa, Lancome and Drakkar Noir, to name a few, have already introduced perfumes and perfume products in India at selective counters. Other well known international brands waiting in the wings to make an entry are Revlon, Christian Dior and Elizabeth Arden.

Major fashion houses in France, sell perfumes and cosmetics carrying their own brand names. The little French town, Grasse, has been the traditional centre for perfume manufacturing of several famous perfumes and the perfume business is the essence and economic mainstay of its people, for which it is known as the world capital of perfumes(Ahluwalia, 1995). Famous among the brands are Channel (Coco Channel, Channel No.5), Christian Dior (Poison), Yves Saint Laurent (Opium), Jean Patoku (Joy), Nina Ricci (La'air Du Temps) and Gucci (Gucci No. 3). Famous personalities from the glitzy world of show-business have often lent their names to promote sales of perfumes such as Zeenat by Zeenat Aman, Picasso by Paloma Picasso, Elizabeth by Elizabeth Taylor and Delon by Alain Delon.

Amongst the Indian manufacturers, the major players in the perfume trade are Helene Curtis (Park Avenue, Premium), Lakme (Shie, Ivana) and Yardley (Chique, Laace). So far, many sub-standard perfumes were being sold in the local shops. Entry of foreign perfumes would upgrade the quality of perfumes in India. Apart from offering a wider range of perfumes, availability of Indian made perfumes of international quality at affordable prices could reduce the demand for smuggled perfumes (at times fakes). Nanz, the super market located in Greater Kailash-I, New Delhi has a very impressive range of Indian made perfumes and perfume products on display (Dey, 1995).

Potential of Traditonal Perfume Industry

Besides kewda, the traditional perfume industries produce attars or absolutes from rose, jasmine, mogra, genda, vetiver, palmarosa and mehindi etc. There is also a flourishing black market of these perfumes to avoid excise or other taxes and no records to verify about the quantum of trade.

The industry has a good potential for growth and would further enhance the potential due to the diversification of perfumes in aromapathy and food flavouring industries. Use of perfumes from plant origin are important in rituals and ceremonies of a society. Salient features of some essential oil bearing plants are given in Table 17, which yield good quality perfumes after processing.

Kewda Perfume Industry

Kewda perfume is obtained by distilling male spadices in water and the *otto* content is 0.03% on weight of spadices (Srivastava and Srivastava, 1976). Its commercial exploitation is centered mainly in Ganjam district of Odisha, accounting for 90% of Indian production of kewda flowers and perfume

(Deshaprabhu, 1965; Dutta *et al.*, 1982). Kotabamali in Tamil Nadu, Jaunpur, Ghazipur and Sahaswan in Uttar Pradesh, Alwar and Pushkar (Ajmer) in Rajasthan are also centres of production (Dutta, 1995). The perfumes are prepared indigenously by the traditional hydro-distillation method (Fig. 11). Although this industry is of centuries old, it came into prominence in the last decade, with a sudden spurt in the consumption of pan masala and chewing tobacco. It has been observed that from a single flower, the industry people of north India manufacture the kewda essence of Rs.25/- and sell the essence @ Rs.3 7.50 to the same people from whom they have purchased @ Rs. 5/- per flower and one can get a profit of Rs. 15000/- from 1000 flowers.

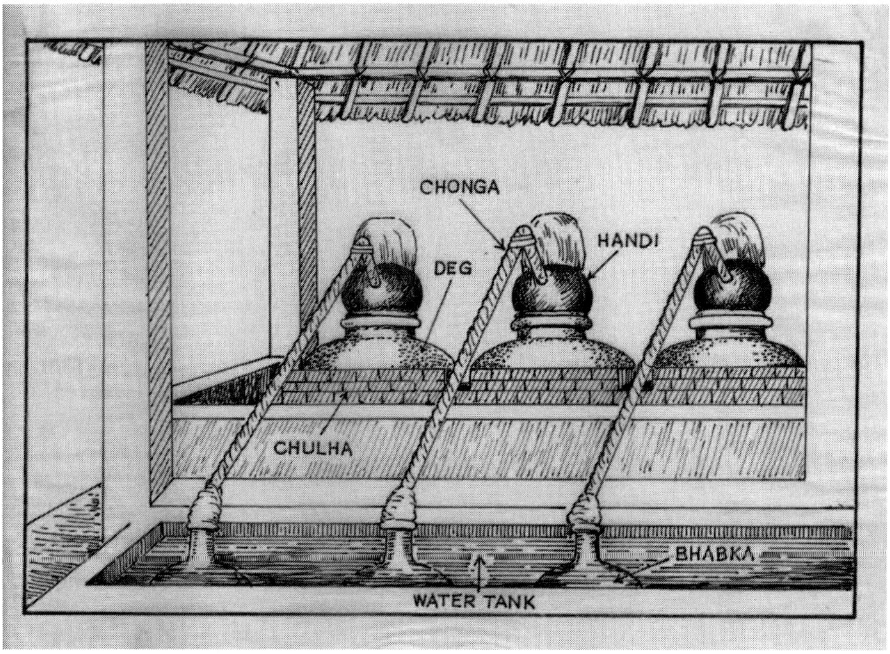

Fig. 11: Distillation of kewda flowers by indigenous method

Historical Background

Kewda distillation industry was first started about 200 years ago in the Berhampur and Chhatrapur regions of Ganjam district by some Muslims from erstwhile Punjab province of undivided India and thereafter some others from Kannauj, Ghazipur, Lucknow, Kanpur and other places of N-India (Sahu and Misra, 2007). Even today, the entire industry still remains so effectively hidden behind the kewda bushes in the country-side that one does not get any trace of it unless the person treks far into the interior away from the National Highway No. 5. The entire industry is monopolized by the people from the north, particularly from Kannauj and some of them stay there throughout the year, while most

Table 17: Salient features of some essential oil bearing plants

Common name	Scientific name	Plant part (s) used	Habitat	Economic stage of plant growth	Harvesting season	Processing method	Yield of essential oil (%)	Uses
Rose	*Rosa damascena* Mill.	Flower	Commercially cultivated mainly in Uttar Pradesh	15 months after cultivation, but maximum flowers obtained till 5th year Carefully maintained plantations remain productive upto 40 years	Early hours of morning in March	Steam distillation soon after harvesting	0.01-0.03	In high grade perfume, largely for flavouring soft drinks, alcoholic liquors and tobacco, particularly snuff and chewing tobacco. It is also used in lotions and eye washes.
Kewda	*Pandanus fascicularis* Lam.	Male spadix	Grows mainly in marshy places. Cultivated in Odisha	Plantations of 15 years and onwards	July-October	Steam distillation	0.01-0.03	Kewda attar and kewda water are popular on festive occasions such as marriages
Palmarosa	*Cymbopogon martini Roxb.*	Upper 2/3rd of the plant	Grows wild in M.P., A.P., Maharashtra, Karnataka and also cultivated in Odisha	At full bloom	Flowering stage	Water and steam distillation	1.20-1.50 (dry weight) 0.25-0.30 (fresh)	In perfumery, soap and cosmetics industry. Also for industrial isolation of geraniol.

(Contd.)

Vetiver (khus)	*Vetiveria zizanioides* (Linn.) Nash	Roots	Grows wild in north India and utilized commercially in Bharatpur (Rajasthan)	15-18 months old plants	Dry period	Water distillation	0.30-0.50	In perfumery and flavouring agents
Vetiver (cultivated)	*Vetiveria zizanioides* (Linn.) Nash	Roots	Cultivated mainly in Kerala	8-19 months old plants	October-December	Water distillation	0.80-1.00	In perfumery and flavouring agents
Jasmine	*Jasminum officinale* L. Var. grandiflorum	Flower	Cultivated mainly in north India and also in Karnataka	1-15 years of plantations	The best yield and quality of perfume is obtained from flowers collected early in the morning	Enfleurage or solvent extraction or steam distillation	0.22-0.26	In perfumery, blends very well practically with any floral scent, leading to smoothness and elegance
Mehendi	*Lawsonia alba* L.	Whole plant	Grown mainly in Haryana, Gujarat and Rajasthan	2nd year onwards	April-May October-November	Steam distillation	0.01-0.02	In perfumery
Sandal wood	*Santalum album Linn.*	Heart wood	Grows wild and also cultivated in Karnataka, Andhra Pradesh, Tamil Nadu and Maharashtra	After full development of heartwood	Mature tree	Steam distillation	4.5-6.0	High grade perfumery and cosmetics.

Source: Anonymous (1973)

others remain only during the season of flower production (April-December) for carrying out the distillation. During 2002, on demand of the kewda growers and farmers union of Ganjam district, Ministry of MSME, Government of India established a project of Fragrance and Flavour Development Centre, Kannauj in the Industrial Estate of Berhampur in the name of Technology Support Centre for Kewda Industry to provide support services to the farmers and now, it has been handed over to a private agency through Kewda farmers union with the support of National Horticultural Mission for management.

Present Set Up

The entire kewda industry in Ganjam district is centered around Chhatrapur, Agraharam, Berhampur, Tulu, Keluapalli, Panapalli and Gopalpur areas. Earlier there were 250 factories, but now there are about 58 factories operating in these areas having each factory with 2 to 30 distillation units (*bhatties*) (Plate 39) and the total number of distillation units operating are about 240 with more than 1000 stills for kewda processing in the district. These units commercially exploit the flowers found in large concentration in the vicinity. Though the kewda plants are also found in other coastal areas of the state, they have not been found suitable for commercial use. An assured supply of 2 lakh flowers per annum is essential for the economic operation of a distillery (Dutta *et al.*, 1987). The distilleries collect flowers from various distances depending on their units. Our study on relationship of extent of utilization of flowers in the distilleries of different units with various dist ances of collection from sea-shore showed that factories with 10 units used flowers from 30 km distance and with 2 units collected flowers from 10 km (Tulu region) to 25 km (Chhatrapur) distance, while factories with 15 or more units collected and utilized flowers from even 40 km distance (Table 18). After the distillation, the distillers supply the products to their own perfumery firms situated most in N-India or they take up orders for attar preparation from various parties, when the sandal wood oil is supplied by the party concerned along with the specifications regarding grade and number of flowers to be distilled. Generally, the representatives of different perfumery houses of N-India used to come during peak season and purchased the products from the factory owners. The factory owners provide all the facilities to the purchasers like free accomodation and visit to nearby places, but food is on payment basis.

Plate 39: A kewda distillery house (*bhatti*)

Table 18: Relationship of extent of utilization of kewda flowers in distilleries of different units with various distances of collection from factory site in Ganjam district

Location	Distance from factory site (km)	Extent of utilization of flowers in distilleries (%)						
		30units	25units	20units	15units	10units	5units	2units
Tulu	5	75.0	80.0	85.0	90.0	95.0	97.0	99.0
	10	7.5	7.5	6.0	5.0	2.5	2.5	1.0
	15	5.5	4.3	3.5	2.0	1.0	0.5	-
	20	4.5	3.6	2.5	1.4	0.8	-	-
	25	3.5	2.5	1.5	0.8	0.4	-	-
	30	2.5	1.3	1.0	0.5	0.3	-	-
	40	1.5	0.5	0.5	0.3	-	-	-
Gopalpur	5	66.0	73.0	72.5	82.0	86.0	90.5	95.0
	10	10.0	8.0	7.5	6.5	5.5	2.5	2.5
	15	8.5	7.0	7.0	5.0	4.5	4.0	1.5
	20	6.5	5.2	6.5	3.5	2.5	2.5	1.0
	25	4.5	3.5	4.0	1.5	1.0	0.5	-
	30	2.5	1.8	1.3	1.0	0.5	-	-
	40	2.0	1.5	1.2	0.5	-	-	-

(Contd.)

Chhatrapur	5	67.0	68.0	72.0	76.0	84.0	88.0	95.0
	10	8.5	7.5	6.0	5.5	4.0	3.0	2.2
	15	6.5	8.5	6.3	4.5	4.0	3.5	1.8
	20	5.5	6.5	5.8	5.5	3.5	2.5	0.5
	25	5.5	4.5	5.1	4.5	2.5	1.5	0.5
	30	3.5	2.8	3.0	2.5	1.2	1.0	-
	40	3.5	2.2	1.8	1.5	0.8	0.5	-

Distillation by Traditional Deg System

Kewda distillation is carried out in the fire heated stills using equipments of traditional type. This age-old method is preferred over the improved ones as it is stated, not only to give better quality of perfume but also cheapest of all other methods. Copper, bamboo, clay and grass are the basic materials required for construction and if a water source is at hand, then a distillery can be set up even in the most rural settings. The distillation owners charge about Rs. 600/- for a batch of 1000 flowers which includes the cost of fuel and labour. At Keluapalli, an indigenous distillation unit was surveyed for its structural details (Table 19).

Method for Kewda Attar Preparation

The fresh flowers are first made free from outer green spinous leaves and spathes if any and the pointed tips of the spathes are broken by hand (Plate 40). Then, a known number of flowers (generally multiples of 500-1000) (Fig. 12) are charged in a big copper still of 60-70 kg weight (locally known as *deg*), kept inside the soil by $3/4^{th}$ of its height over an open furnace (*chulha*) (Plate 41) and water is poured @ 60 litres per 1000 flowers or 2:1 ratio of flowers to water by volume (Plate 42). Generally, mats are often kept in the still under the flowers to prevent charring. The capacity of the *deg* is more than 200 litres and can contain about 1200 flowers. An earthen pot or a copper lid (*handi*) inverted over the still is connected to a copper receiver vessel (*bhabka*) through a bent bamboo pipe (*chonga*); which is covered spirally and tied with rope outside and acts as a *condenser*. The capacity of the *bhabka* is 20 litres. During the preparation of attar, normally 5 kg of sandal wood oil is kept in the *bhabka* to absorb the aroma of kewda flowers. The *bhabka* is kept cool by submerging it in a shallow cement tank filled with water. All joints or connections are made air-tightened by sealing with the help of cloth and a paste of white clay (*multani mitti*). The *deg* is heated gently from beneath with ordinary wood fire over the *chulha* and on heating, the kewda vapours mixed with steam pass through the bamboo connector and get absorbed over the sandal wood oil before condensing in the *bhabka*. The speed of distillation is controlled by regulating the intensity of fire. The kewda aroma is thus gradually absorbed in the sandal wood oil to make the kewda attar (Plate 43).

Table 19: Structural details of an indigenous distillation unit of kewda

Sl.No.	Name of the part	Making material	Capacity (litre)	Height (cm)	Circumference (cm)	Diameter at middle (cm)	Neck height (cm)	Neck diameter (cm)	Neck Circumference (cm)
1.	Deg (Copper still)	Copper	250 (Copper used = 90 kg)	75	300	100	15	40	120
2.	Bhabka (receiver vessel)	Copper	25	60	120	40 (inner)	30	10	40(at base)
3.	Handi (earthen pot)	Soil	20	50	135	42	10	30	90
4.	Chonga (Condenser)	Bamboo Covered Spirally With rope	Length of short bent pipe (cm): 30.0 Length of long pipe (cm):150.0 Diameter of hole (cm): 2.5						

Distillation is continued for 4-6 hours. Generally, during the distillation of 6 hours, three receiver vessels are used. First receiver vessel containing 5 kg of sandal wood oil is disconnected after initial 2 hours and likewise, the second vessel containing the sandal wood oil is disconnected at the end of 4 hours of distillation and during the last 2 hours of distillation, the third empty vessel without containing sandal wood oil collects only the condensed scented water. In each 2 hours of distillation, about 5 litres of condensed water is collected in the vessel besides the absorption of aroma by the sandal wood oil. During the distillation process, water in the cement tank is also changed alternatively to keep the water cool (Plate 44). Then the vessels are disconnected and corked and kept overnight for setting (Plate 45). Next morning, the condensed water forming the lower layer is separated out from the vessels through a small outlet provided at the bottom (Plate 46) and the water is recirculated in the next charge. This process is repeated over a number of days (normally 5-7 days) allowing the same sandal wood oil to absorb the aroma until the specified number of flowers are distilled depending on the grades of attar. After the distillation, the bhabka is allowed to cool and left untouched for 1-2 days and then the attar is separated from water by the help of a separating funnel and the water is reused for distillation, so that perfume in the water is not lost. Finally, the attars are packed in air tight galvanised iron containers (Plate 47 and 48) and transported to different centres of perfume industries situated at Kannauj, Lucknow, Kanpur and other places of North India for blending and manufacture of various preparations (Fig. 13). The distilleries remain active from 8.00 AM to 10.00 PM of the day for the processing of 8000 to 40,000 flowers per day per distillery (Hota, 2012).

Plate 40: Removal of outer green spinous leaves/spathes of the flowers tomake ready for distillation (www.whitelotusaromatics.com)

Fig. 12: Kewda flowers being loaded in a deg for distillation

Plate 41: Copper stills (*degs*) in the kewda distillery

Plate 42: Water being poured into the tank to start the distillation process.

Plate 43: Kewda distillation in progress

Plate 44: Completion of the kewda distillation

Plate 45: Corked receiver copper vessels (bhabka) with kewda attar

Plate 46: Separation of kewda water from the vessel through the bottom outlet

Plate 47: Kewda attar/water poured into and sealing of the GI containers

Plate 48: Kewda attar/water stored in the GI containers for transport

Fig. 13: Kewda attar/water packed in G.I. drums being loaded on a truck for transport.

The final processing of kewda perfumes is kept a trade secret by the perfumery houses. After finishing, the ready perfumery products are sent for sale in the national and international markets. Sometimes, refined liquid paraffin, commercially known as white oil (locally called as colol), is used in place of sandal wood oil. Liquid paraffin is not only cheaper but also absorbs the aroma mor quickly, thus involving less cost in the preparation of attar. However, due to the poor power of retention of aroma, liquid paraffin is used only when a cheaper grade is required. Also at times, the users, especially the manufacturers of some tobacco preparations give specifications for blending palmarosa oil, which is mixed @ 10 g per kg of sandal wood oil before starting the distillation. Addition of palmarosa oil is believed to improve the quality of attar. Several grades of attar are prepared depending on the number of flowers distilled and aroma absorbed per kg of sandal wood oil, which normally varies from 1000 to 10,000 and known as *ek hazari* to *dus hazari* grades, respectively. The specifications regarding number of flowers to be consumed and grade of attar to be prepared are given by the perfumers or the users. No standard other than the aroma is known for determining the quality of attar.

Method for Kewda Water Preparation

Kewda water is prepared by distilling the flowers with water in the similar manner as of kewda attar without taking the sandal wood or any base oil in the *bhabka* and collecting the distillate in an empty receiver vessel. Here, the water is added to the flowers @ 100 litres per 1000 flowers. One male labourer is sufficient for the process. Kewda water is designated as *ek hazara, do hazara, panch hazara* and *dus hazara*, when 18 liters of distillate are collected by distilling 1,000, 2,000, 5,000 and 10,000 kewda flowers, respectively. The initial 18 litres of water distilled is locally known as *agadi* grade and the subsequent 18 litres of water distilled from the same flowers is known as *pichhadi* grade. About 10,000 litres of floral water are annually produced in the country (Deshprabhu, 1965). The kewda water is kept in galvanised iron containers and the sealed (Fig. 14). The expiry date of the kewda water remains for 2 years from the date of sealing. The cost of 18 litres of kewda water comes to Rs.3,000/-

Fig. 14: Kewda attar/water stored in G.I. drums at the distillery house

Method for Kewda Oil Preparation

Fresh flowers collected before 9.00 a.m. are utilized for extraction of kewda oil (*rooh* kewda). The recovery of the volatile oil from the flowers collected after 9.00 a.m. is very less. The distillation process is similar to the attar extraction without taking the base oil in the *bhabka*. The oil collected remained floating over the condensed water and the oil is decanted and filtered and then processed like attar. The recovery of oil is around 15-20 g per 1000 flowers. Generally, kewda oil is not extracted on a commercial scale due to its high solubility in water (Ambasta, 1986). Hence, for obtaining '*rooh kewda*', skill is required to control on fire beside quantity and rate of distillation to ward off loss in distillate. A minimum of 35,000 flowers are required to produce one kg of oil, sometimes which may go upto one lakh flowers. Kewda oil has been experimentally prepared by extracting the flowers with a solvent, precipitating the fatty and waxy matters with alcohol and distilling the absolute under reduced pressure (Deshaprabhu, 1966, Dutta *et al.,*1987).

Sometimes perfume of the kewda spadices is extracted with sesame seeds by the modified enfleurage process, as used for jasmine and the extracted oil is used as hair oil alone or after admixture with jasmine-perfumed sesame oil. (Deshprabhu, 1966).

However, improvement in the collection system and means of transport from remote kewda growing areas to the distillation centres may promote the industry. Improvement of the existing age-old methods of distillation may also increase the recovery of oil and quality of perfumes.

Trends in Kewda Industry

In earlier years, kewda essence was not traded in its absolute oil form, instead it was used to be absorbed into sandal wood oil. In the trade, the quality of attar is measured on the basis of number of flowers distilled and absorbed into a unit volume of sandal wood oil, but there is no fool proof method to assess the actual number of flowers distilled and absorbed. Consequently, the trade is going on in good faith, which is also common in other perfume trades like rose, mogra and genda etc. With the increasing price and demand for these oils, malpractices have become rampant and the actual quantities never matched with the claims. Hence, now there is a clear trend in shifting of demand from attars to absolutes. The quality of absolute can be assessed by GLC (Gas Liquid Chromatography) method, for which the industries both domestic and foreign prefer to buy absolute. The absolute is diluted by mixing with sandal wood oil by the user industries themselves.

Utilization and Oil Recoveryof Flowers

The investigation on the mode of utilization of flowers in a distillery and oil recovery per 1000 flowers by indigenous distillation method at 3 locations of Ganjam district (Table 20) revealed that maximum number of flowers were utilized (95 %) for perfume extraction during the rainy season (June-October) and rest 5% during other months. More than 50% of the flowers collected were used for preparation of kewda attar during January-May and for oil during June-October and about 25% of the flowers were utilized for kewda water production in rest of the months. The oil recovery was better during rainy season (June-October) at Tulu as compared to Keluapalli and Panapalli areas. The colour of the oil is golden-yellow during winter season (November-January) and light yellow to yellow in rainy season. Kewda attar is produced in large scale, while production of kewda water remained with a few units in Chhatrapur block. It is also interesting to note that the units located in Rangeilunda block produce most of the '*rooh*', while attars are produced mostly by Chhatarapur units (Rao *et al.,* 1996).

Generation of Man-Days

Unemployment and under-employment are the major causes of rural poverty in Odisha. Small, marginal and landless families are mainly dependant on agriculture for their livelihood. In drought prone areas, agriculture can provide employment for hardly 100 days in a year and such seasonal employment can not sustain them throughout the year. The employment kewda generates utilizing the wasteland in the rural areas and the secondary income it brings to the rural poor can not be ignored.

A kewda distillery in its smooth operation has generated a number of economic man-days. A standard kewda distillery having 15 units generates about 160 man-days per day during the month of September followed by 61 man-days per day during August (Table 21) (Maharana *et al.,* 1993) in its different operations like collection, counting and processing of flowers. The collection of kewda flowers provides an additional income to the rural poor as the plucking and delivery of flowers to the distilleries are over during the morning hours by 9.00 a.m. without disturbing the main vocation of the people.

On an average, a kewda distillery remains in operation economically for 200-250 days a year and a minimum of 5 casual labourers are engaged to get their wages @ 250 to 300/- per day. It is interesting to note that till date there is no labourer strike demanding higher wages in any of the distilleries. The state industry department should encourage the establishment of distillery units in self-employment scheme and the economic condition of the rural poor kewda growers can be improved by formation of a kewda corporation. Hence, conscious cultivation of kewda deserves promotion by various aencies instead of leaving it to grow naturally.

Table 20: Month-wise utilization and oil recovery of kewda flowers in a distillery at different locations of Ganjam district

Month	Total number of flowers utilised in a distillery at Keluapalli	Extent of utilization of flowers (%) at Tulu			Kewda oil recovery (g/100 flowers)			Colour of oil
		Kewda oil	Kewda attar	Kewda water	Kelua-palli	Tulu	Pana-palli	
July	13,950	50	25	25	17.5	18.4	17.2	Light yellow
August	6,20,000	85	10	5	18.5	19.5	17.9	Yellow
September	7,50,000	90	10	-	20.2	21.8	19.8	Yellow
October	46,500	50	15	35	24.6	25.2	23.6	Yellow
November	13,500	40	35	25	10.8	11.6	10.5	Golden yellow
December	13,175	30	45	25	15.5	16.2	14.8	Golden yellow
January	6,200	20	55	25	10.2	11.3	10.1	Golden yellow
February	4,200	10	70	20	NA*	NA	NA	-
March	7,750	10	75	15	NA	NA	NA	-
April	13,500	10	75	15	NA	NA	NA	-
May	15,500	10	75	15	15.5	16.2	15.0	Light yellow
June	15,000	50	30	20	16.8	17.6	16.3	Light yellow

Table 21: Economic man-days generated per day in various operations of a standard kewda distillery

Season	Month	No. of flowers collected	Labour engaged for collection			Mode of transportation (%)			No. of labourers counting of flowers	Labourers for processing			Total economic man-days generated
			Male	Female	Child	Head	Cycle	Other vehicle		Male	Female	Child	
Rainy	Jul	2,400	6 (3)	12 (6)	24 (6)	60	25	15	1	2	2	2 (1)	21.0
	Aug	6,825	18 (9)	36 (18)	72(18)	50	30	20	2	10	3	2 (1)	61.0
	Sep	20,550	55 (27.5)	110 (55)	220 (55)	40	30	30	4	15	3	2 (1)	160.5
	Oct	1,500	4 (2)	8 (4)	16 (4)	70	20	10	2	3	1	2 (1)	17.0
Winter	Nov	1,000	3 (1.5)	6 (3)	12 (3)	80	15	5	2	3	2	2 (1)	15.5
	Dec	500	2 (1)	4 (2)	8 (2)	85	10	5	1	2	1	2 (1)	10.0
	Jan	400	1 (0.5)	2 (1)	4 (1)	90	10	-	1	2	1	2 (1)	7.5
	Feb	350	1 (0.5)	2 (1)	4 (1)	90	10	-	1	2	1	2 (1)	7.5
Summer	Mar	300	1 (0.5)	2 (1)	4 (1)	90	10	-	1	2	1	2 (1)	7.5
	Apr	250	1 (0.5)	2 (1)	4 (1)	90	10	-	1	2	1	2 (1)	7.5
	May	300	1 (0.5)	2 (1)	4 (1)	90	10	-	1	2	1	2 (1)	7.5
	Jun	400	1 (0.5)	2 (1)	4 (1)	75	20	5	1	2	1	2 (1)	7.5

N.B: Equivalent man-days are given in parentheses. *As collection is over by 9.00 am, 1 labourer = half man-day.

13

Marketing and Trade

The world trade in flavour and fragrances in 1987 was US $ 7,000 million and at present, it is estimated at more than $ 10,000 million annually. In 1975, perfume was used by one woman in ten, but today it is used by eight out of ten women; even reticent Japanese women are adopting the habit (Anonymous, 1995 c). The average annual industry growth rate in perfumes of India is valued at about US $ 20 million with an annual growth rate of 12% (Murty, 1993),while the UK market is Rs. 2000 crores and the US market supersedes it by Rs. 4000 crores (Walia, 1997). However, the domestic luxury perfume market is estimated to grow at a rate of 20-25 % (Biswas, 2012). Besides domestic market, India earns a lot of foreign exchange from the export of total essential oils and perfumes and is fast growing as it is showed from the trends in export earnings of Rs. 11.1 corers in 1982-83 rose to 4-fold in 1986-87 of Rs. 41.1 crores (CHEMEXCIL, 1988). According to a market survey report, Ahmed (1999) reported that the daily turn over of the attar market in India is more than 50 crores.

Middle East countries continue to be the big buyers of our traditional perfumes. Kewda flavour which is used in pan masala and tobacco is a familiar one in this part of the world. Kewda essence, being one of the few flavouring oils, has a good scope for export to other developed countries where the food flavouring industry is well developed. India is the leading producer of kewda perfumes in the world and the market for kewda perfumes is fast expanding specifically in the Arab countries. India exported the kewda water of Rs. 266.2 lakhs in 1986-87 which rose to Rs. 276.4 lakhs in 1987-88 (CHEMEXCIL, 1990). Anonymous (2007) reported that about 140 distilleries (bhattis), both local and outsider, earn about Rs.1.3 crore by way of extraction charges and annually they extract 332 kgs of kewda oil (Rs.8.3 crore), 7823 kgs of kewda attar (Rs.15.65 crores) and 150 MTs of kewda water (Rs.4.47 crore), the total coming to around Rs.28 crore. For extracting attar , the extractors use about Rs. 8 crore worth of sandal wood oil procured mainly from other states, while Rs.18 crore goes to the plantation owners as flower cost and to the flower collectors.

The total trade is worth Rs. 20 crore annually for the local people from whom perfume traders from different parts of the country purchase basic kewda products. After value addition in their units, they sell perfumes and fragrances making hefty profits. However, despite kewda's potential as a livelihood resource, hardly any attention has ever been paid to its scientific cultivation or extraction with an eye to business. Kewda perfumery trade is controlled by distillers of Ghazipur, Kannauj, Lucknow and neighbouring parts of north India.

Kannauj is considered as the perfume capital of India, where 80% of the perfume trade is linked with pan masala and chewing tobacco industries (Anonymous, 1996). The volume of business in this trade is to the tune of Rs. 500-800 crores, of which pan parag alone shares about 50% and rest 50% is shared by the brands like Baba jarda, Prabhat jarda, Gopal jarda, Rajanigandha, Radha, Ashiqui and Kesar etc. In this industry, of the total perfumes used, kewda alone accounts for about 75% by value (Rao, 1994), hence the importance of kewda is amply clear in the perfume trade.

The cost of production of pan parag is about Rs. 3500/- per kg, of which Rs. 1000/- is spent towards perfumes and kewda is the major one, which accounts for about Rs. 750/-. Assuming the cost of production of pan masala is 35% of its sale price, the turn over of the kewda is of the order of Rs. 450-550 crores. This industry is growing at an annual rate of 15-20%; hence the demand for kewda perfumes is also expected to grow at a minimum level of 15-20 %. Panigrahy (2019) reported that Ganjam district has processed around 3500 tonnes of kewda flowers of worth about Rs.10 crores.

Production Cost of Kewda Attar

Production cost of kewda attar at the factory depends on the quality of sandal wood oil used and the number of flowers distilled. Generally, attar is stored in 5 kg capacity galvanised iron cans. With the prevailing rate of distillation carges and price of flowers, the working cost of kewda attar is estimated below (Table 22).

Table 22: Estimatation of cost of preparation of kewda attar
Kewda Attar (*Ek hazara* = 1000 flowers)

Sl.No.	Items	Amount (Rs.)
1.	Cost of sandalwood oil (1 kg)	20,000.00
2.	Price of 1000 flowers @ Rs. 8/-	8,000.00
3.	Distillation charge @ Rs. 600 per 1000 flowers	6,00.00
4.	GST @ 18%	4,968.00
	Production cost of kewda attar per kg	33,568.00

Although the cost of kewda attar works out to be Rs. 33,568 per kg, the prevailing market price is Rs. 25,000/- to 30,000/- per kg. The discrepancy may be due to the highly secretive trade and evasion of tax, which seems to the order of the day. The number of flowers actually distilled is also far less than the quoted figure as there is no system of measuring the exact number of flowers distilled and in this way, the price of Rs. 25,000/- to 30,000/- is becoming economical. The same also explains the price of absolute (*rooh* kewda) at about Rs. 3.0 lakhs per kg (Anonymous, 1999). About 35,000 flowers are necessary for production of a kg of absolute, in which the flowers alone cost more than Rs. 2.5 lakh; hence the possibility of adulteration of the product and evasionof tax due to the secretive trade can not be ruled out (Rao, 1994).

Marketing System

The marketing system in perfume industry is well developed as the money flows from top to bottom. But in case of kewda distillation industry, the flower producers are poor small farmers and scattered over a large area. The collection agents play a decisive role and charge handsomely due to the perishable nature of the flowers. The marketing channels followed in the industry are almost similar to the trades of other perfumery crops like rose *(Rosa damascena),* bela *(Jasminum sambac)*, genda *(Tagetes erecta)* and khus *(Vetiveria zizanioides)* etc. (Fig. 15).

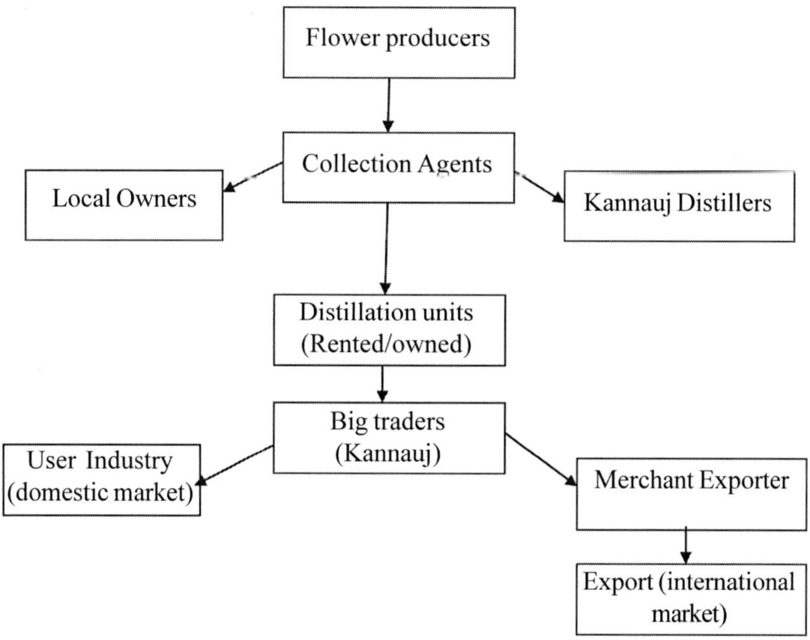

Fig. 15: Marketing channels in kewda perfume industry

The marketing system of Kewda perfume industry has various components like flower growers, collection agents, distillers, traders, user industries and merchant exporters etc.

In the process, the growers or farmers plant kewda on the farm bunds and collect flowers on maturity by themselves or give the plants on lease annually to other farmers. They also collect the flowers from the natural plantations. For supply of the entire flowers harvested, they take advance from the collection agents at a predetermined price and on terms and conditions fixed in advance. The collection agents act as middle men between the growers and the distillers and they receive the advanced amount from the distillers to pass on to the growers for ensuring a regular supply of flowers to the distillers. The distillers produce the essential oil by their own distillation units or on rented units. However, most of the distillation units are owned by the local entrepreneurs or entrepreneurs of Kannauj, who settled in Ganjam district since generations. The traders from Kannauj procure the entire kewda oil produced in the area and they usually give credit to the local entrepreneurs. Practically, they control the entire trade because of their money power and established trade links with the user industries, which include mostly the pan masala and scented tobacco industries. In the trade of kewda industry, the buyers are limited and the suppliers are in large numbers; hence it seems to be a buyers market despite the short supply of the product. Then for international market, the export of the products is done through the merchant exporters.

Trade Practices

As per Rao (1994), the procurement procedure by the user industry is to call for tenders and the competitive bidder is awarded with the contract for the year. The price of the oil is fixed based on the prevailing price of the flower in Ganjam district and supply is made on monthly basis. The payment is made 60-90 days after delivery of the consignment.

Small quantities of kewda attar or absolute are also being exported abroad through the exporters from Kannauj, Delhi and Mumbai. For exports, the merchant exporters contact the dealers of the importing countries and send the sample along with a quoted price. Once the sample is approved, the price is negotiated and the schedule of the supply is fixed. The merchant exporters procure the consignments from the distillers or traders and export the same acting as middle men or commission agents. Sometimes the big traders with their own distillation units pool the perfumes from small distillers and directly contact the importers abroad and execute the orders for export to international market.

Price Margins at Various Levels

The flower growers are very poor farmers. The margins on investment or price spread at different levels are assessed through discussions with the farmers and traders, although it is very difficult to arrive at exact figure. However, the following figure can be the best indicative (Rao, 1994) (Fig. 16).

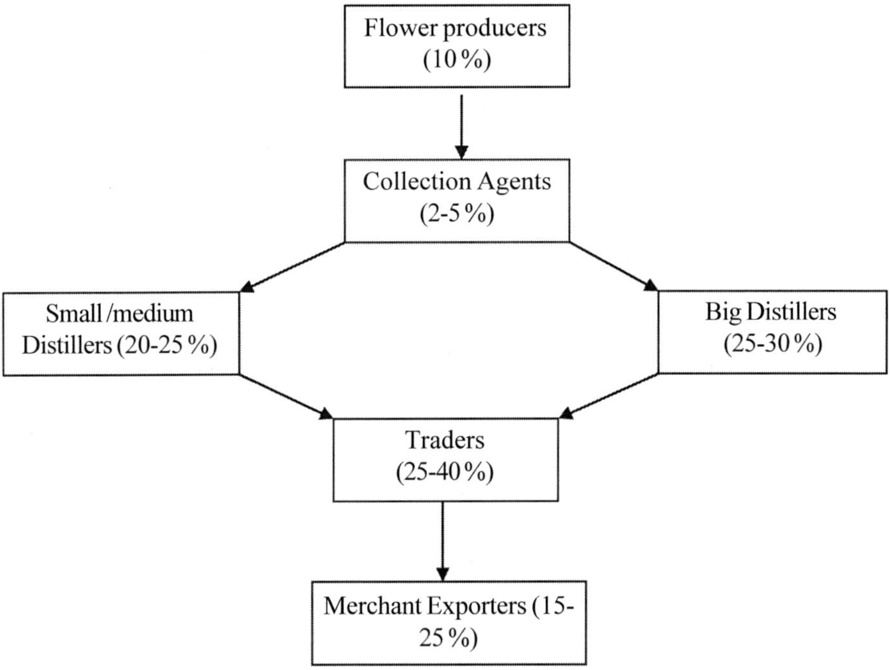

Fig. 16: Spread of margins on investment/price at different levels of business

Hence, from the above, it is indicated that the traders get maximum margins in the business.

14

Economics of Cultivation

Kewda plants are mostly propagated through vegetative means. The common propagating materials used are ground suckers. The planting materials are planted in paired-row method maintaining a spacing of 4m from plant to plant and row to row keeping a spacing of 6 m between paired rows. In a field of one hectare, a minimum plant population of 500 plants in the field and 150 plants on the field bunds with a total of 650 plants are maintained. A fertilizer dose of 30:15:15 kg NPK/ha and 4 cartloads (100 cft) of FYM/ha are applied. The planting of collected ground suckers is normally done in the field just before monsoon after thorough land preparation under rainfed condition.The earthing-up of the plants are done up to five years of planting for quick initiation of flowering at the age of 5 years after planting. During the first year, the operational cost per hectare comes to Rs.28,900/- and from 2nd to 5th years, it comes to Rs.6,600 per year and then the maintenance cost from 6th year onwards comes to Rs.6,000/- per year. The details of the operational cost per hectare of kewda cultivation under rainfed condition of coastal Ganjam district is given at Table 23. Kewda plant comes to flower on the 5th year of planting, but the annual net profit begins on the 6th year of planting and the cumulative net profit starts on the 10th year (Table 24). The economic viability of kewda plantations is observed from 10th to 30th year of planting and after 30th year, the net profit declines. A farmer can earn an annual profit of Rs.61,000/- per hectare of kewda cultivation from 11th to 15th year, which goes on increasing and from 26th to 30th year, the annual net return reached to Rs.1,23,400/- with a cumulative return of Rs.4,04,970/- on the 30th year of planting. Then at the age of 35th year, it declines to Rs.1,07,800/- and at 40th year to Rs.87,000/- per annum (Table 24). Although the profit was calculated only on the basis of sale of flowers yield, some additional profit can be obtained from other plant parts like leaves and stilt roots. After distillation of one lakh flowers, the spent over flowers and plant parts can be utilized as fertilizer/manure by selling at a rate of Rs.1000/- per batch. In the two blocks (viz., Chhatrapur and Rangeilunda) of Ganjam district, about 7,000 women are engaged for collection of flowers, leaves and stilt roots and they prepare ropes and other household items from the roots and leaves. It has been observed that one female can prepare 12 bundles of ropes each of 12 m length and the sale rate of rope is about Rs.10/- per metre. The ropes are sent to the markets of Visakhapatnam (Andhra Pradesh) and Chennai (Tamil Nadu) for sale. The profit can also be increased by incorporating some inter-crops like *Casuarina*, coconut, Agasti and medicinal crops like Patchouli etc. Further, kewda can be taken on the sand dunes and waste saline land of the coastal lines, where no other crops can be grown profitably. Hence, kewda needs better attention for its extensive cultivation with a foresight on the perfume industry.

Table 23: Details of operational cost per hectare in kewda cultivation under rainfed condition

Age of planting	Item	Quantity	Rate (Rs.)	Amount (Rs.)	
1st year	i) Land preparation:				
	Ploughing 3 times with desi plough (2 pairs for 3 days)	6 ploughs	500	3000	
	Levelling and clod breaking	2 ladders	500	1000	
		2 CLs	280	560	
	Lay out and digging of pits	10 CLs	280	2800	
	ii) Manuring and fertilization (just before monsoon)				
	Applicationof FYM	4 cart loads (100 Cft)	450	1800	
	Application of fertilizers @ 30:15:15 kg NPK/ha	Urea 65 kg	276/50kg	360	2840
		SSP 94 kg	362/50kg	680	
		MOP 25 kg	802/50kg	400	
		5 CLs	280	1400	
	iii) Planting suckers (just before monsoon)				
	Collection and planting of ground suckers	30 CLs	280	8400	
	Fungicide (Bavistin) treatment of suckers	2.5 g/100 suckers	170/100g	170	
	iv) Maintenance, watch and ward	3 CLs	280	8400	
			Total	28920	
			Say	28900	
2nd year to 5th year	i) Fertilization and earthing up (just before monsoon)				
	Applicationof fertilizers	Same as 1st year		2840	
	Earthing up	10 CLs	280	2800	
	Plant protection measures	2 sprayings	500	1000	
			Total	6640	
			Say	@6600	
6th year onwards	i) Maintenance cost of the plants (i.e., fertilization, plant protection, desuckering and plucking of flowers etc, as and when necessary)	-	6000	@6000	

CL = Casual labourer

Table 24: Economics of kewda cultivation per hectare (planting method: Paired-row; spacing: (4m x 4m) x 6m, Plant population : 650)

					Anticipated					
Age of planting (year)	Annual operational cost (Rs.)	Interest on operational cost @ 10 % (Rs.)	Annual expenditure (Rs.)	Cumulative expenditure (Rs.)	Flower production/ plant	Flower production/ ha	Annual return @ Rs.8/ flower	Cumulative return (Rs.)	Annual net profit (Rs.)	Cumulative net profit (Rs.)
1	28900	2890	31790	31790	-	-	-	-	-31790	-31790
2	6600	660	7260	39050	-	-	-	-	-7260	-39050
3	6600	660	7260	46310	-	-	-	-	-7260	-46310
4	6600	660	7260	53570	-	-	-	-	-7260	-53570
5	6600	660	7260	60830	1	650	5200	5200	-2060	-55630
6	6000	600	6600	67430	2	1300	10400	15600	+3800	-51830
7	6000	600	6600	74030	3	1950	15600	31200	+9000	-42830
8	6000	600	6600	80630	4	2600	20800	52000	+14200	-28630
9	6000	600	6600	87230	6	3900	31200	83200	+24600	-4030
10	6000	600	6600	93830	9	5850	46800	130000	+40200	+36170
11-15	6000	600	6600	100430	13	8450	67600	197600	+61000	+97170
16-20	6000	600	6600	107030	17	11050	88400	286000	+81800	+178970
21-25	6000	600	6600	113630	21	13650	109200	395200	+102600	+281570
26-30	6000	600	6600	120230	25	16250	130000	525200	+123400	+404970
31-35	6000	600	6600	126830	22	14300	114400	639600	+107800	+512770
36-40	6000	600	6600	133430	18	11700	93600	733200	+87000	+599770

15

Economic Uses

1. House Plant

The screwpines or kewda plants are characteristics of many tropical regions with long ringed trunks, bracing roots and crowns of *Dracaena*-like foliage. The young plants are amongst the most popular foliage plants for home decoration as the plants do not flower at early stages (Wickham, 1982a). They are especially suited for fern-pans, table decoration and grown to a very large extent by wholesale florists and palm specialists (Bailey, 1950). They are also grown as pot and tub specimens for green house, residence, veranda and lawn decoration where their stiff clustered foliage gives them a formal decorative character. They may be kept as indoor ornamental foliage plants in the warm and semi light places of a house. (Wickham, 1982a).

2. Fence

The plant is useful for binding the soil, so checks soil erosion and not browsed by the cattle due to its spiny leaves; hence sometimes used as fence around the crop field in the coastal districts of Odisha, Andhra Pradesh, Tamil Nadu, Kerala and Gujrat. The plants are often used for hedges as in the *Casuarina* plantations at Puri of Odisha (Haines, 1961a) (Fig. 17).

Fig. 17: Kewda planted with *Casuariana* on sea shore to check soil erosion

3. Fibre

The leaves composed of tough, longitudinal, white, glossy fibres, which are employed for covering huts, making mats, cordage and in South India for larger kinds of hunting nets and drag-ropes of large fishing nets. The excellent mats are made out of it and used by the rich and poor alike in Kerala in large scale for sleeping purpose (Pandey, 1984). In Mauritius, the fibres are employed to make sacks for coffee, sugar and grain. The leaves are cut in every second year and each plant yields sufficient fibre to make two large bags. In Solomon Islands, the leaves are used for manufacture of protective covers, locally known as custom umbrellas (Rickard and Cox, 1984).

In Philippines and other Pacific Islands, the leaves are cured either by heating or sun drying, with or without prior soaking in water. Superior quality mats are said to be made from young leaves freshly harvested, rather than from dead leaves gathered in the ground. The leaves are said to be a good paper making material (Deshaprabhu, 1966).

The roots (Plate 49) are also fibrous and are used by the basket-makers for binding. When cut into lengths and beaten out, commonly used as brushes for painting and white washing. The root fibre may be used as a substitute for bristle in brush making. In the Ganjam district of Odisha, the aerial stilt roots are generally collected by low income group people at the end of rainy season (i.e from September onwards) and are processed by cleaning and soaking in water for few hours and then dried under the sun. The recovery of dried fibres is about 15% of the fresh roots by weight (Table 23). The clean dried fibres are sold to the middlemen @ Rs. 1250-1300 per quintal, which then exported to the neighbouring states like Andhra Pradesh, Madhya Pradesh, Tamil Nadu and Karnatak by trucks for industrial use. Sometimes, people sell the fibres in the local markets by making heap of the fibres. The dried root fibres are used for making mats (Plate 50), ropes for fishing lunch and nets, packing materials for fruits and vegetables, binding the cattle, binding material for fencing and thatching (Plate 51) the huts and thick fibres as brush for white washing of the walls of pucca house. The study on mechanical strength of fibres reveal that water soaked fibres are stronger than fresh and dried fibres (Table 25). Hence, to develop rural economy, small cottage industries can be set up to utilize the different plant parts of kewda for making various domestic fancy items.

Table 25: Growth characters and breaking stress of aerial stilt root of kewda

Location	Length (cm)	Girth (cm)	Diameter (cm)		Weight (g)	
			Fresh	Dry	Fresh	Dry
Rangeilunda	200	15.0	5.0	3.5	1750	250
Gopalpur	195	14.8	4.2	3.0	1600	200
Indrakhi	180	13.5	4.5	3.4	1650	229

Location	Length of fibre (cm)			Breaking stress (kg)*		
	Fresh	Dry	Water soaked	Fresh	Dry	Water soaked
Rangeilunda	30	20	20	2.50	2.95	4.00
Gopalpur	30	20	20	2.40	3.06	4.50
Indrakhi	30	20	20	2.15	3.50	5.20

* by young's apparatus

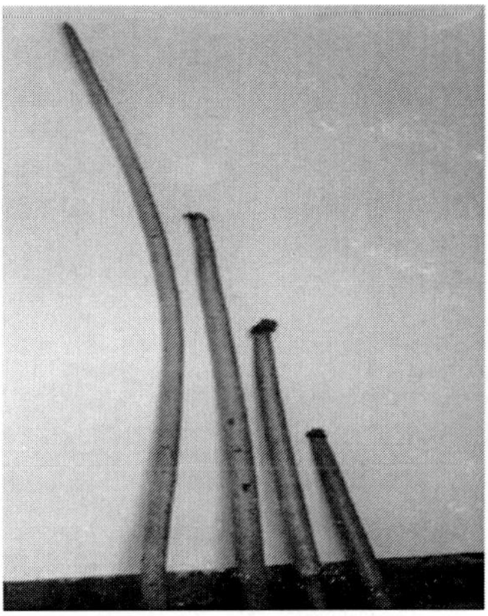

Plate 49: Collected kewda stilt roots

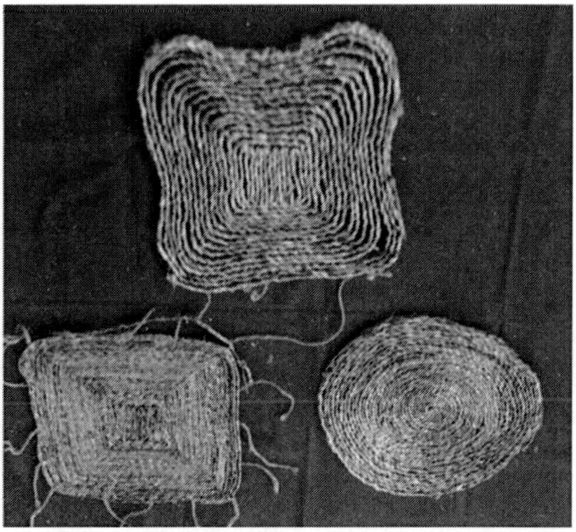

Plate 50: Mats made from kewda stilt root fibre

Plate 51: Ropes and binders made from kewda stilt root fibre

4. Food

In Philippines, the terminal bud and the tender floral bracts are eaten raw or cooked with various condiments. The pulp contained in the lower part of the drupe is eaten in parts of Southern India at times of scarcity although they contain considerable amount of raphides or crystals that cause irritation in the mouth, however, there are distinct races which contain so few raphides that they serve as staple food in parts of Polynesia. A race of this species, known in the Marshall Island is said to have a sweet flesh with the flavour of apple. The seeds are also edible, but their separation from surrounding pulp is laborious (Deshaprabhu, 1966). The lower pulpy part of the drupes is eaten by the natives of the Peninsula in times of scarcity (Royle, 1970) and when ripe, is sweet smelling and orange-red in colour (Plate 52 and 53). In the Gilbert Islands and elsewhere in the Pacific, the flesh is mixed with coconut milk or grated coconut and the paste baked into thin flat cakes, which may be powdered for storage or used in preparation of other dishes (Purseglove, 1975). The investigation on food value of kewda fruit showed that a ripe fruit contains moisture: 80%, crude protein: 0.8-1.0%, pulp: 16-18%, total soluble solids: 10-15%, acidity: 0.4-0.7% and reducing sugar: 0.14% (Fig. 18) and the fleshy base of the drupes can be preserved as squash, jelly and slices in syrup (jam) (Plate 54). The flowers also together with catechu and certain spices, form a substance called *kia khoir* is used in betel (pan) (Watt, 1892).

Plate 52: Kewda female flower at different stages of development

Plate 53: A fruit bearing kewda plant

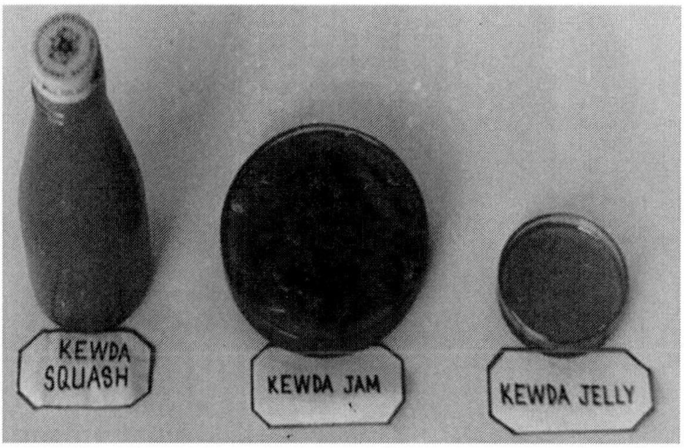

Plate 54: Squash, jam and jelley prepared from kewda fruit

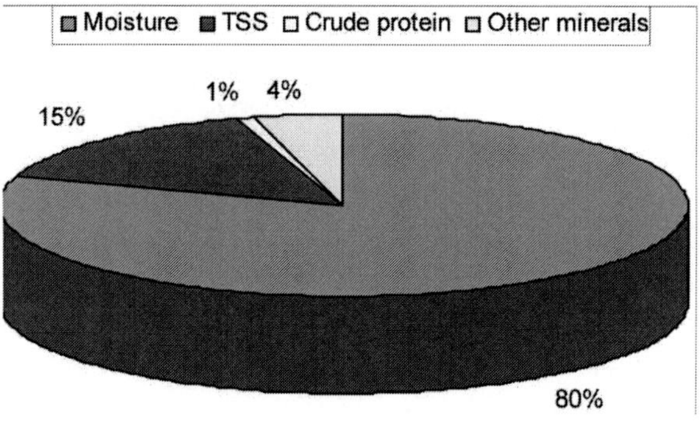

Fig. 18: Food value of kewda fruit

5. Perfume

The white spathes of the male inflorescence emit a fragrant smell, which is much valued in India and Burma and used for hair decoration of ladies, flavouring catechu and kept in boxes to lend a lasting fragrance to the stored garments. Kewda attar, water and oil (*rooh kewda*) are prepared from the spathes by steam distillation or in the more general sense, by hydro-distillation. Kewda attar is one of the most popular perfumes in India, where it has been used since ancient times. It blends well with almost all types of fancy perfumes and is used for scenting clothes, flavouring betel (*pan masala*) and tobacco, bouquets, lotions, cosmetics, soaps, hair oils, tobacco and agarbatti (Plate 55). On festive occasions, kewda water is used for flavouring varoius foods, sweets, syrups and soft drinks, which are especially popular in North India. In Arab countries, kewda water is used during bathing.

Plate 55: Different kewda products like incense sticks, kewda water and attar

6. Domestic

In Ganjam district, generally, the leaves are used locally as fuel, but of late, they are utilized for making several fancy articles like vanity bags, designed mats, money purse, flower baskets and hats etc (Plates 56-58). To increase the strength and quality of colour, the leaves are processed before use (Plate 59). The process involves splitting of middle aged leaves to remove the marginal prickles and the spiny mid-rib. Then the splitted leaves are boiled in water for a period of one and half hours followed by steeping in cow milk for half an hour and then dried under the sun. After drying, the leaves are strewed to different sizes and length depending on the purposes and designs to be used. In South India, the leaves are used for making umbrellas and in Kerala, the leaves are bleached and cut into pieces for making excellent mats (Plate 60), which are used by rich and poor alike in large scale for sleeping purpose (Pandey, 1984). The thick aerial rope-like roots are also cut into lengths and beaten out at one end to form brushes and are also employed as a substitute for corks. Demand for kewda leaf products has increased due to their nice look and durability. The eco-friendly, bio-degradable products have found wide acceptance among the city dwellers also (Anonymous, 2011a). During summer days, the plants having low or no flower production are cut and dried up by the villagers to use those as fuelwood (Plate 61).

Plate 56: Circular mats, tumbler cover and money purse made from kewda leaves

Plate 57: Designed asan and flower basket made from kewda leaves

Plate 58: Designed mats made from kewda leaves

Plate 59: Processed kewda leaves ready for making usable items

Plate 60: Folding mat made from kewda leaves

Plate 61: Kewda plants cut by the villagers for fuel purpose

7. Sacred

Generally, the flowers are worn to the hair by the girls to win the love of their lovers. Legends also reveal that Parvati approached Shiva with kewda flowers, which attracted Shiva's attention and he got disturbed in his penance. Another legend says that this plant is cursed by Lord Shiva to be thorny for telling a lie and giving false evidence on the occasion of a dispute between Brahma and Vishnu, but he (Shiva) again took pity on it and order that the flower would be highly aromatic and he (Shiva) should be worshipped with the flowers of this plant on the day of Shiva Ratri (Plate 62). The flowers are also employed in the worship of many other gods (Watt, 1892). Till today kewda flowers are offered to Lord Jagannath, even some matthas in Puri are maintaining kewda plantations in their campus. A good kewda and ketaki plantation is existing near Puri at Balanga.

Plate 62: Kewda flowers being sold at lingaraj temple, Bhubaneswar on Shivaratri (Prameya Odia daily: Feb. 26, 2014)

8. Medicine

The leaves are pungent, bitter, with flavour and said to be valuable in leprosy, small-pox, scabies, syphilis, leucoderma and diseases of heart and brain, tumours and aphrodisiac (Deshprabhu, 1966, Kirtikar and Basu, 1935). The root (2 g ground in cow milk) taken for 7 days prevents abortion (Anonymous, 2014a). The aerial root is also used medicinally by the Sinhalese (Watt, 1892). The anthers of the male flowers (5-10 g) are given in earache, headache, leucoderma, eruptions and diseases of blood. The juice obtained from the whole inflorescence (spathes removed) in til *(Sesamum indicum*) oil (1-2 tsp or 250-300 mg) massage once daily is useful in rheumatic arthritis in animals (Anonymous, 2014a). The oil and otto (attar) are considered stimulant, antispasmodic and are administered for headache and rheumatism and the doses of distillate *(arka*) and syrup (*sharbet*) are 50-70 ml and 28-56 ml, respectively. In Yunani medicine, the oil cools and strengthens the brain, in Ayurveda, the fruit is useful for bronchytis, rheumatism and urinary discharges and in Cambodia, the root is considered as diuretic, depurative and tonic (Kirtikar and Basu, 1935, Deshaprabhu, 1966, Kapoor, 1990). Kewda attar can cure both high and low blood pressure if used on a constant basis. Sharbat keora is just the thing to beat the rising mercury (Ahmed, 1999).

References

Ahluwalia, I.R. 1995. Fragrant grasse. Saturaday Times. *The Times of India*. New Delhi.:7.

Ahmed, F. B. 1999. Scent percent Indian. Sunday magazine. *The Hindusthan Times* (English Daily), New Delhi. Apr. 25: 3.

Ambasta, S. P. 1986. *Pandanus*. In. The useful plants of India. C.S.I.R., New Delhi : 423-424.

Anonymous, 2011. Cultivation of kewda. *Kalinga Putra* (Odia Weekly). (15-21 Nov.): 8.

Anonymous, 2014a. *Medicinal use of kewda. Prameya (Odia daily)*. (25 Aug.): 15.

Anonymous. 1973. Classification of essential oil bearing plant materials. In: Indian Standard Classification of Essential Oil Bearing Aromatic Plants. Indian Std. Institution, New Delhi (Nov. Issue): 9-13

Anonymous. 1988. General agricultural characteristics of the state. In: Status Report, OUAT Reg. Res. Stn., Ranital, Balasore: 1-2.

Anonymous. 1992. Application of essential oils. Food Talk. *35th Ann. Con. Essen. Oil Assocn. India*, Bombay. **5**(2) : 2.

Anonymous. 1995. Cursed ketaki, *Sachitra Vijaya* (Odia Monthly). Dolton Publ., Madras, October issue. 2(1): 11-12.

Anonymous. 1995a. Annual export of ketaki flowers from Orissa accounting to Rs. 50 crores. *The Prajatantra* (Odia Daily), Cuttack. Sept. 26,pp.1.

Anonymous. 1995b. French perfumers seek fragrance. *The Times of India* (English Daily), New Delhi. (Dec. 28). 158 (288):13.

Anonymous. 1995c. Heaven scent. In: Saturday times, *The Times of India* (English Daily), New Delhi. (Dec. 23). 158 (284):6.

Anonymous. 1996. Kewda thorn in the flesh of steel plant. *The Hindustan Times* (English Daily), New Delhi, (Jun. 21). 72 (171): 5.

Anonymous. 1999. Super cyclone devastated 5000 acres of Kewda Plantations and the life line of lakhs of families is at stake. *Sambad* (Odia Daily), Nov. 22:6.

Anonymous. 2007. Kewda can attract big bucks. India First (English weekly) (14 Jul.): 5.

Anonymous. 2008. Mahak Madak (Odia). Sambad Sunday Magazine (31 Aug.): 3

Anonymous. 2014. State sketches plans to boost agriculture exports. *The New Indian Exp.*, Bhubaneswar Edn. (10 Nov.): 4.

Anonymous.2011a. Kewda leaf products get a new avatar. The new Indian Express. Bhubaneswar Edn. (22 Jul.): 5.

Ash, J. 1987. Demography, dispersal and production of *Pandanus tectorius* (*Pandanaceae*) in Fiji, *Australian J. Bot*. **35**: 313-330.

Bailey, L. H. 1950. *Pandanus.* Ini. Std. Cyclopedia Hort., Macmillan Co., New York, 3: 2449-2451.

Beura, S. and Jagadev, P. N. 1994. Post-harvest care and utilization of cashew apple. *Souvenir*. Orissa Univ. Agric. Tech., Bhubaneswar. (Apr. 5): 28-30.

Bhat, K. G 1992. Studies on screwpines (*Pandanus*) of Karnataka-new records. Indian *J. Forestry*. **15**: 359-360.

Bhattacharjee, Sivakali. 1987. Ketaki. In: Chiranjeevi Banausadhy. Anand Publ. Calcutta. pp.35-42.

Bisht, M.; Sharma, S.; Mathela, C. S. and Maheshwari, M. L. 1997. Composition of essential oil from the flowers of 'Kewda' (*Pandanus odoratissimus* L.). *Indian Perfumer.* **41** (2) : 69-72.

Biswas, V.S. 2012. Masters of fragrances. The New Indian Express, Edex (9 Jan.) : 12

Burkill, I. H. 1966. A dictionary of economic products of the Malay Peninsula. Min. Agric. Co-op., Kualalampur.

Chai, M.; Soepadmo, E. and Jong, H.S. 1979. Proceeding of workshop on genetic resources of plants, animals and micro-organisms in Malaysia. *Malaysian Appl. Biol.*, Spl. Issue. **8**:1.

Chandra, G. 1985. Investigations on essential oils and isolates of potential value at HBTI, Kanpur. *Indian Perfumer*. **29** :23-30.

Chavan, P. B. and Hosagoudar, V. B. 1984. Three new fungi from Satara, Maharastra. *Indian. J. Econ. Taxon. Bot.* **5**: 447-450.

CHEMEXCIL. 1988. Trends of exports 1982-83 and 1986-87. Jhansi Castle 4th floor, 7 Cooperage Road, Bombay.

CHEMEXCIL. 1990. Trends of exports 1986-87, 1987-88 and 1988-89. Jhansi Castle. 4th Floor, 7 Cooperage Road, Bombay.

Chittenden, F.J. 1956. Pandanaceae. In: Royal Hort. Soc. dictionary of gardening. Oxford at Clarendon Press, U.K. (2nd edn.), 3: 1476-1477.

Chopra, R. N.; Nayar, S. L. and Chopra, I. C. 1956. *Pandanus* (Pandanaceae). Glossary of Indian Medicinal Plants CSIR, New Delhi :184.

Cox, P. A. 1990. Pollination and the evolution of breeding systems in *Pandanaceae. Annals of Missouri Bot. Gdn.* **77**: 816-840.

Deshaprabhu, S. B. 1965. Perfume industry. In: The Wealth of India : Industrial Products. CSIR, New Delhi. **6** : 223-226.

Deshaprabhu, S. B. 1966. *Pandanus*. In: The Wealth of India : Raw Materials. CSIR, New Delhi. 7 : 216-221.

Deshpande, S.S. 1938. Essential oil from flowers of kewda *(Pandanus odoratissimus)*. *Indian J. Chem. Soc.* : 509.

Dev, S. 1993. Indian essential oils and aroma chemicals Response feature. The *Times of India* (English Daily) New Delhi (Feb. 11):12.

Dey, M. 1995. A whiff of the West. Saturday Times. *The Times of India* (English Daily).: 7.

Dhingra, D. R.; Gupta, G. N.; Jain C. and Shukla, U. N. 1951. Otto of *Pandanus odoratissimus* L. or kewda. *Perfum Essential Oil Rec.* **42**: 114-117.

Dhingra, S. N.; Shukla, U. N.; Gupta, G. N. and Dhingra, D. R. 1954. Essential oil of kewda (*Pandanus odoratissimus* Linn. f.). *Perfum. Essential Oil Rec.* **45**: 219-222.

Dutta, P. K. 1995. Kewda. In: Advances in Horticulture – Medicinal and Aromatic Plants (Chadha, K.L. and Gupta, Rajendra, Eds.) Malhotra Publ. House, New Delhi. 11:833-841.

Dutta, P. K.; Saxena, H. O.; Mishra, B. C.; Paul, S. C.; Brahmam, M.; Sahoo, S. and Patra, P. 1985. Cultivation and survey of essential oil bearing plants in Orissa region. *Indian Perfumer.* **29**:63-66.

Dutta, P. K.; Saxena, H.O. and Brahmam, M. 1982. Kewda distillation industry in Orissa. In. Cultivn. & Utilizn. of Aromatic Plants. RRL (CSIR), Jammu-Tawi: 634-639.

Dutta, P.K.; Saxena, H.O. and Brahmam, M.1987. Kewda perfume industry in India. *Econ. Bot.* **41**:403-410.

Everette, T. H. 1969. Angiosperms : the monocotyledons. In: The Living Trees of the world. *Thames & Hudson Ltd.*, London : 66-68.

Farooqi, A.A. and Sreeramu, B.S. 2001. History, present status and future prospects of essential oil crops. In: Cultivation of Medicinal and Aromatic Crops. Univ. Press (India) Ltd., Hyderabad: 265-279.

Gandhi, M. 1989. Screwpine. In: Brahma's hair: mythology of India Plants. Rupa Publn. , New Delhi. :121-124.

Ghosh, S.K. 2001. Basmati-Pandanus: as aromatic ad basmati: Agric. Industry Survey (Aug. issue): 18.

Graf, A.B. 1986. Pandanus. In: Tropica: *Color cyclopedia of exotic plants and trees* (3rd edn.). Roehrs Co. Publ. New Jersey, USA: 1054.

Haines, H. H. 1961. Pandanus. In: The Botany of Bihar and Orissa. *Bot. Survey of India*, Calcutta. (Reprinted Edn) 3: 918-919.

Haines, H. H. 1961a. Pandanus. In: The Botany of Bihar and Orissa. *Bot. Survey of India, Calcultta*. (Reprinted Edn.) 2: 717.

Heyne, K. 1927. *Die Nuttige Planten van Naderlandsch Indie*. Vol.1. Department van Landbouw, Nijverheid & Handel, Buitenzorg :127.

Heywood, V.H. 1978. Monocotyledon families.In: Flowering Plants of the world. Oxford Univ., Press, London : 305-306.

Hota, J. 2012. High valued kewda flower- a story (Odia). Dharitri Sunday Magazine (22-28 Jan.): 3-4.

Hyndman, D.C. 1984. Ethnobotany of Wopkaimin *Pandanus* : significant Papua New Guinea plant resource. *Econ. Bot*. **38** : 287-303.

Jagadev, P. N.; Naik, B.; Sahoo, S. C.; Beura, S. and Maharana, T. 1993a. Variability, selection and planting material in kewda (*Pandanus fascicularis* Lam.). *Indian J. Genet*. **53**: 310-314.

Jagadev, P.N. Beura, S. and Maharana, T. 2001. Insect pests and diseases of kewda (Pandanus fascicularis Lam.). *Indian Perfumer*. **45**(4): 265-268.

Jagadev, P.N., Beura, S., Naik, B. and Maharana, T. 1999 & 2000. Promotion of flowering by planofix in kewda. *Curr. Agric. Res.* **12 & 13**(1-4): 46-47.

Jagadev, P.N., Sahoo, S.C., Beura, S., Naik, B. and Maharana, T. 1992. Kewda: a crop for wasteland utilization and development of rural economy in Orissa. Proc. Workshop on Sustainable Horticulture, Bhubaneswar (22 Feb.): 94-98.

Jagadev, P.N.; Beura, S. and Maharana, T. 1993a. Prospects of aromatic plants in Orissa. *Orissa J. Hort*. **21**: 95-98.

Jagadev, P.N.; Naik, B.; Beura, S. and Maharana, T. 1993. Correlation and regression study of flower production and its components in kewda (*Pandanus fascicularis* Lam.) *Indian J. Genet*.**53**: 315-317.

Jong, K. 1979. Biological aspects of plant genetic resource conservation in S-E Asia. *Instt. S-E Asian Biol.*, Univ. Aberdeen, Scotland, U.K.

Kapoor, H.N. 1979. Commercial Utilization of keora (*Pandanus tectorius)* flowers Symp. on Prodn. and Utilizn. of Forest Pdts. RRL (CSIR), Jammu-Tawi. **1** : 10-11.

Kapoor, L. D. 1990. *Pandanus tectorius* Soland. ex Parkinson. In: Handbook of Ayurvedic Medicinal Plants. CRC Press, Florida, USA.pp.255.

Katiyar, S. K.; Kumar, N. and Bhatia, A. K. 1989. A chemical study of *Pandanus lerum* fruit grown in the Andaman and Nicober Islands. *Trop. Sci*.**29** : 137-140.

Kirtikar, K. R. and Basu, B. D. 1935. Pandanaceae. Indian Medicinal Plants. L. M. Basu Publ., Allahabad. (2nd edn.). **4**: 2591-2593.

Kumar, P.N. 1995. Perfume merchants of golden flowers. Sachitra Vijaya (Odia Monthly). Doltorn Publ., Madras. October issue. **2**(1): 12-13.

Lewis, F. 1934. Vegetable products of Ceylone. *Assoc. Newspapers of Ceylone Ltd.*, Colombo: 365.

Lorus and Milne, M. 1967. Cattails, burreeds and screwpines. In:Living plants of the world. Chanticleer Press, New York :270-272.

Macoboy, S. 1986. *Pandanus*. In: What Tree is That ? Tiger Books Intl. Ltd., London:158.

Maharana, T. and Jagadev, P. N. 1993. Final report of ICAR project on productivity and industrial use study in medicinal and aromatic plants of Orissa with special reference to *Pandanus* and *Cymbopogon* OUAT, Bhubaneswar : 51p.

Maharana, T., Jagadev, P.N., Beura, S.a nd Naik, B. 2001. Selection strategy for high flower production in kewda. Proc. Natl. Seminar on conservation and utilization of medicinal and aromatic plants (4-6 Dec.), RRL (CSIR), Bhubaneswar: 251-255.

Maharana, T.; Jagadev, P. N.; Naik, B. and Beura, S. 1995. Plant growth and flower production in kewda under different edapho-climatic conditions. Proc. Natl. Symp. on Adv. in R & D in Hort. for Export. CCSHAU, Hisar. (Jan 28-30):89.

Maharana, T.; Jagadev, P.N.; Naik, B. and Beura, S. 1993. Kewda perfume industry and rural employment in Orissa. *Indian Perfumer*. **37**:24-28.

Maheshwari, M. L. 1995. Composition of essential oil from flowers of keora (*Pandanus odoratissimus* Linn.) by capillary gas chromatography. *Indian Perfumer.* **39** (1) :45-48.

Mahmud, J.1992. On a scented route. **In:** Sunday Review, The Times of India (English Daily), New Delhi. (Sep.13). pp.4.

Manral, K. 1996. Scents and sensibility. The Sunday Rev. *The Times* of India (English Daily). Dec. 1: 5.

Massal, E. and Barrau, J. 1956. Food plants of the south sea islands. Noumea : South Pacific Commission, Tech. Paper No. 94.

Misra, R. and Rao, Y.R. 1997. A GC-MS Study of kewda (*Pandanus odoratissimus* Linn.) oil. *Indian Perfumer.* **41** (4) :143-145.

Misra, Reeta; Dash, P.K. and Rao, Y.R. 2000. Chemical composition of the essential oils of kewda and ketaki. *J. Essential Oil Res.* **12**(2) :175-178.

Mohammad, S. 2000. Kewda loses its incense. In: Region Scan. *The Hindustan Times* (English Daily) Jan 5, pp.1.

Mohammad, S. 2000. Kewda loses its incense. *The Hindustan Times* (Region Scan). Jan. 5, pp.1.

Mott, R.C. 1975. Foliage plants. In: The complete book of house plants. Octopus Books Ltd., London : 73.

Murty, K.S.S. 1993. Flavour industry in India. Response Feature. *The Times of India* (English Daily), New Delhi. (Feb.11). pp. 11.

Myers, N. 1983. A wealth of wild species. West-view Press. Boulder, Colorado, USA.

Naqvi, A. A.; Mandal, S. and Kumar, S. 1995. Quality parameters of keora *Pandanus odoratissimus* essential oils. *Curr. Res. Med. Arom. Plants*. **17**: 338.

Nigam, M.C. and Ahmad, A. 1992. Chemical and gas chromatographic examination of essential oil of *Pandanus odoratissimus* (keora) flowers. *Indian Perfumer*. **36**:93-95.

Oldfield, M. L. 1984. The value of conserving genetic resources. Natl. Parks Service, U.S. Dept. Interior, Washington DC, USA.

Padhi, S.K. 2017. Kewda flower industry in distress. Dharitree (Odia daily) (14 Aug.): 8.

Panda, K.K., Mohapatra, S., Das, L.N., Misra, M.K. and Panda, B.B. 2000. Optimal utilization of kewda (Pandanus fascicularis) to ameliorate economy and e3cology of coastal India. Proc. Natl. Seminar on Frontiers Res. Dev. Med. Plants, CIMAP, Lucknow: 679.

Pandey, B. P. 1984. Fibres and fibre yielding plants In : Economic Botany S. Chand & Co. Ltd., New Delhi. (3rd Edn) : 163.

Panigrahy, S. 2019. Kewra industry dying a slow death in Ganjam. The New Indian Exp., Bhubaneswar (20 Jan.): 5.

Patnaik, Naveen. 1993. Fragrant screwpine (*Pandanus odoratissimus*). In: The Garden of life. Aquarian Harper Collino Publn., London, pp.154.

Platt, B. S. 1962. Tables of representative values of foods commonly used in tropical countries. London: HMSO.

Purseglove, J. W. 1975. Pandanaceae. In: Tropical Crops :Monocotyledons ELBS Longman Group Ltd., London, U.K. (1st edn.), : 511-515.

Randhawa, G. S. and Mukhopadhyay, A. 1986. Ornamental foliage plants. In: Floriculture in India. Allied Publ. Pvt. Ltd., New Delhi : 266-267.

Rao, M. S. 1994. Study of perfume industry with special reference to kewda in Ganjam district, Orissa. NABARD, Bhubaneswar. 63p.

Rao, Y. R.; Misra, R., Misra, B. C., Panigrahi, M. R. and Ray, H. S. 1996. Whither "Kewda" *(Pandanus odoratissimus)* perfume industry *Indian Perfumer*, **40** (3) :95-98.

Rayaguru, K., Sahoo, G.R., Khan, Md. K. and Das, A.K. 2008. Pandanus amayllifolius: a potential substitute of basmati aroma. Souvenir and Abstracts. *Natl. Seminar on 'Commercial exploitation of medicinal and aromatic plants for health and sustenance'* (Feb. 9-10), OUAT, Bhubaneswar; 106.

Rendle, A. B. 1979. Pandanales. In: The classification of flowering plants (gymnosperms and monocotyledons). Vikas Publ. House Pvt. Ltd., Ghaziabad, U.P. (2nd edn.) **1** : 187-197.

Rickard; P. P. and Cox, P. A. 1984. Custom Umbrellas (poro) from Pandanus in Solomon Islands. *Econ. Bot.* **38**: 314-321.

Royle, J.F. 1970. Pandanaceae. In: Illustrations of botany and of Himalayan mountains and flora of Cashmere. Today &Tomorrow's, Printers & Publ., New Delhi (Reprinted Edn) 1:408-409.

Sachdeva, G. 1993. The Indian perfumer from attars to perfumes. Response feature. *The Times of India* (English Daily), New Delhi, (Feb. 11):10.

Saha, B.M. and Sareen, M. 1997. Open space. In: The Sunday Times of India, New Delhi, Aug. 17: 24.

Sahay, V. 1998. Fragrant faith. In: *Sunday Magazine*. The Hindustan Times (English Daily), New Delhi, Jul., 26, pp.7.

Sahu, D. and Misra, M.K. 2007. Ecology and traditional technology of screwpine perfume industry in coastal Orissa. *Indian J. Traditional Knowledge*. **6**(2): 253-261.

Singh, D.B. 1997. *Pandanus* : a versatile tree for Andaman tribes. *Indian Hort.* **42** (1): 8-9.

Singh, U.; Wadhwani, A. M. and Johri, B. M. 1983. Dictionary of economic plants in India. I.C.A.R., New Delhi: 163.

Southwell, K. and Harris, R. 1992. Chemical characteristics of *Pandanus conoideus* fruit lipid. *J. Sci. Food Agric.* **58** : 593-594.

Srivastava, A. and Srivastava, R. 1976. Essential oil industry in India. *Indian Perfumer*. **20**: 99-113.

St. John, H.1965. Revision of the genus *Pandanus* Stickman: Additional Malayan species of *Pandanus.Pacific Sci.* **19** :224-237.

Stone, B. C. 1976. Pandanaceae. **In:** Flora of Hassan district, Karnataka. (Saldanha, C. J. and Nicolson, D. H. eds.). *Amerind Publ. Co.,* New Delhi : 779-781.

Stone, B. C. 1978. Studies in Malesian Pandanaceae XVII on the taxonomy of 'Pandan Wangi' a *Pandanus* cultivar with scented leaves. *Econ. Bot.* **32** : 285-293.

Subudhi, D. 2012. Sad story of kewda farmers, *Odia Weekly* 365: 25-26.

Subudhi, D. 2013. Arab also flavoured with Ganjam Kewda Flowers. Sambad Sunday Magazine (Odia) (28 Apr.): 4.

Suryanarayanan, T. S. and Narasimham, C. L. 1981. A leaf blight of *Pandanus odoratissimus* caused by *Botryodiplodia theobromae Indian Phytopath*. **34** :254-256.

Tanaka, T. 1976. *Pandanus*. **In:** Tanaka's Cyclopedia of edible plants of the world. Keigaku Publ. Co., Tokyo, Japan: 520-522.

Teng, L. C.; Shen, T. C. and Goh, S. H. 1979. The flavouring compound of the leaves of *Pandanus amaryllifolius. Econ. Bot.* **33**: 72-74.

Thomas, A. 1965. Plants with noble foliage and water plants. **In:** Gardening in Hot. Countries. Faber & Faber Ltd., 24, Russel Square, London. 112-113.

Tomlinson, D. L. 1988. A leaf and fruit disease of *Pandanus conoideus* caused by *Erwinia carotovora* subsp. *Carotovora* in Papua New Guinea. *J. Phytopath.* **121** : 19-25.

Walia, N. 1997. On a scented trail. In : Sunday Times. *The Times of India* (English Daily), New Delhi & Lucknow. Aug. 9. pp.1

Watt, G. 1892. Pandanus. In: A dictionary of the economic products of India. W.H. Allen & Co., London. 6(1): 4-6.

Wickham, C. 1982. *Pandanus* (screwpine). Dictionary of foliage plants. *The House Plant Book.* William Collins Sons & Co. Ltd., London :134.

Wickham, C. 1982a. Where to place your plants. **In:***Dictionary of foliage plants*. In: *The House Plant Book*. William Collins Sons & Co. Ltd., London : 62-63.

Willis, J. C. 1960. *Pandanus*. In: A dictionary of flowering plants and ferns. *The Univ, Press, Cambridge.* (6th edn.) : 485-486.

Colour Plates

Chapter 1: Introduction

Plate 1: Thick kewda plantations along roadsides and paddy field bunds in Ganjam district.

Plate 2: Kewda plants planted with *Casuarina* to check soil erosion and as fence to the main crop in Rangeilunda area

Chapter 3: Botany and Nomenclature

Plate 3: Freycinetia arborea (Fiji) (Graf, 1986)

Plate 4: Freycinetia funicularis male inflorescence (Java, Indonesia) (Graf, 1986)

Plate 5: Freycinetia multiflora (Phillipines) Climbing pandanus (Graf, 1986)

Plate 6: Freycinetia rigidifolia (Thailand, Malaya, Borneo) (Graf, 1986)

Plate 7: Freycinetia stenophylla (New Guinea) (Graf, 1986)

Plate 8: Pandanus tectorius (New Guinea) (Graf, 1986)

Plate 9: Kewda (*P. fascicularis*) Plant

Plate 10: Ketaki (*P. odoratissimus*) Plant

Plate 11: Annapurna (*P. latifolius*) Plant

Plate 12: *Ketaki (P. odoratissimus) Flower*

Plate 13: *Kewda (P. fascicularis)* male flower

Chapter 4: Species and Varieties

Plate 14: *Pandanus leram (Nicober breadfruit)* (Graf, 1986)

Plate 15: *Pandanus utills (Madagascar)* (Graf, 1986)

Plate 16: *Pandanus veitchii (Polynesia)* (Graf, 1986)

Plate 17: *Pandanus baptistii (Melanesia)* (Graf, 1986)

Plate 18:. *Pandanus brosimus (New Guinea)* (Graf, 1986)

Plate 19: *Pandanus pristis (Madagascar)* (Graf, 1986)

Plate 20: *Pandanus rockii (Manoa Valley)* (Graf, 1986)

Plate 21: *Pandanus sanderi Roehrsianus (Malayasia)* (Graf, 1986)

Chapter 5: Breeding and Crop Improvement

Plate 22: A dark green large thorny type kewda plant

Plate 23: A light green large thorny type kewda plant

Plate 24: A dark green small thorny type kewda plant

Plate 25: A thornless type kewda plant

Plate 26: A mala type kewda plant

Chapter 7: Propagation

Plate 27: A kewda seedling

Plate 28: Propagation by kewda stem cutting

Plate 29: Kewda research field at Konark

Plate 30: Propagation by kewda ground sucker

Plate 31: Propagation by kewda crown from flowering part

Chapter 9: Harvesting and Yield of Flower

Plate 32: Kewda plants in full flowering stage

Plate 33: Kewda male flower at different stages of development

Plate 34: Kewda flower collection with the help of a stick (www.whitelotusaromatics.com)

Plate 35: Kewda flowers being carried over head for distillation (www.whitelotusaromatics.com)

Chapter 11: Insect Pests and Diseases

Plate 36: Kewda plants attacked by bag worm (*Manatha albipes*)

Plate 37: Foot rot disease of central shoot in kewda

Plate 38: Leaf blight disease of kewda

Index

Symbols

A

B

C

D

E

F

Q

R

S

T